Einstein's Universe

Also by Nigel Calder

Nigel Calder

Einstein's Universe

 The Viking Press : New York

Published in 1979 by The Viking Press
625 Madison Avenue, New York, N.Y. 10022

LIBRARY OF CONGRESS CATALOGING IN PUBLICATION DATA

Calder, Nigel.
  Einstein's universe.
  Includes index.
  1. Relativity (Physics). 2. Astrophysics.
3. Cosmology. I. Title.
QC173.55.C34   521   78-26087
ISBN 0-670-29076-9

Printed in the United States of America
Set in Video Compano

Third printing May 1979

Author's Note

'How should we celebrate Einstein's centenary?' When that question came from BBC television and the American PBS network, I was in no doubt about my answer: 'Let's make relativity plain.' To do it properly, and in a modern way, required the direct consultation of some seventy physicists and astronomers in thirty institutions on both sides of the Atlantic. They were, as always, unstinting in their help, and I regret that I can mention only a few of them: Kenneth Brecher, Sidney Drell, Roger Penrose, Wallace Sargent, Dennis Sciama, Irwin Shapiro, and John Archibald Wheeler, who assisted all the way through to their participation in our main film-making. Any defects are my fault, of course, not theirs.

Although it accompanies the television film of the same title, this book is separately conceived and written, to convey the ideas and experiments in printed words rather than with moving pictures. I am grateful to the BBC for the double challenge and for the facilities for extensive travels; also for unwavering support for the project.

By always demanding clarification of key ideas, Martin Freeth, the producer, and Peter Campbell, of BBC Publications, held me to that promise about making relativity plain.

The television special *Einstein's Universe*, marking the centenary of Albert Einstein's birth, was made by BBC-TV as a coproduction with WGBH (Boston). Peter Ustinov introduced and narrated the programme, which was largely filmed at the McDonald Observatory, in Texas, with the cooperation of its director, Harlan Smith.

Martin Freeth produced and directed the programme, assisted by John Dollar. Henry Farrar was the cameraman and Alan Jeapes devised the animations. The designer was Colin Lowrey, and David Havard prepared the visual effects. Christopher Woolley edited the film, and Nigel Calder wrote the script.

Contents

Illustrations follow page 73.

Einstein's Universe

1 : Cosmic Whirlwind

Gravity and high-speed motion affect your view of the world.
General Relativity deals with gravity.
Special Relativity deals with high-speed motion.
Appearances change but the laws of physics survive.
The search for a black hole shows Einstein's ideas in action.

Galaxies, stars, planets and now spaceships rush about the universe, and we have a sense of time passing because the positions of objects change. For example, the Sun seems to march across the sky every day, because the Earth itself is spinning. It is upon such changes in the scenery, near and far, that human ideas of space, time and motion depend.

Albert Einstein revolutionised those ideas of space, time and motion. In doing so he solved several cosmic mysteries, and rewrote the theory of gravity. His equations anticipated our age of nuclear energy, space-flight, radar, laser beams and atomic clocks. Einstein pointed to pos-sibilities, so far unfulfilled, of transcending time. And he laid down rules governing the history and fate of the entire universe. The geneticist J. B. S. Haldane called Einstein 'the greatest Jew since Jesus'. But my purpose here is explanation, not hero-worship. Einstein often wondered why he was adulated by people who had not taken the trouble to find out what he had actually said. 'Are they crazy or am I?' he asked.

The photographs showing Albert Einstein as a wrinkled old man, with hair like a superannuated sheepdog's, do not present the real Einstein, but a gentle ghost who haunted Berlin and Princeton, con-demned to the treadmill of a futile project for unifying electricity and gravity, until his death in 1955. The Einstein honoured by later genera-tions expired long before—in 1919, say, when he became a celebrity at the age of forty. By then his great work was finished. Between 1904 and 1917, or between his twenty-fifth birthday and his thirty-eighth, he had rebuilt the universe in his head. The landmarks of Einstein's work are Special Relativity (1905), which deals with high-speed motion, and General Relativity (1915), which deals with gravity.

In what follows I shall take advantage of the fact that more than sixty years have passed since Einstein completed his theory. All subsequent investigation has shown that we actually live in a universe very like that which he described and has confirmed and developed many of the ideas latent in his equations. I mean therefore to sketch Einstein's universe in an up-to-date fashion, with no apology and only a few historical asides. In the conventional, chronological approach the reader is too often lost in the boondocks of Special Relativity, confused by flashing lights and stopwatches, and barely able to glimpse the great ideas of rest-energy and einsteinian gravity. For many years, General Relativity was regarded as too difficult and opaque for ordinary scientists, never mind the man in the street. With hindsight it seems that a generation of mathematical physicists had no vivid grasp of the theory themselves. Now all that is changed. In the era of atomic clocks and suspected black holes it is both possible and appropriate to devote the greater part of this book to Einstein's theory of gravity.

Einstein is often said to have held that 'all things are relative'. He did not. 'Relativity' is in fact a thoroughly bad name for the theory: Einstein considered calling it the opposite: 'invariance theory'. He discovered what was 'absolute' and reliable *despite* the apparent confusions, illusions and contradictions produced by relative motions or the action of gravity. The chief merit of the name 'relativity' is in reminding us that a scientist is unavoidably a participant in the system he is studying. Einstein gave 'the observer' his proper status in modern science.

By embarking upon relativity he set stars and imaginary vehicles reeling like phantoms in the whirlwind he had raised. When the clocks went awry and distances shrank, how were space and time to be recorded reliably? What was real and what was illusory? How could you describe an object and the gravity at work upon it, when the energy of the object depended on who was looking at it? In short, would the laws of nature be the same for everyone, regardless of his place and motion? Perhaps: Einstein vacillated about this proposition, sometimes courting it, sometimes rejecting it. In finding that it was correct, he arrived at the vision that this book sets out to describe, on the hundredth anniversary of Einstein's birth.

I am helped in my task by the fact that relativity became 'hot science' in the 1970s. As recently as 1962 the physicist and populariser of science George Gamow could write, with accuracy:

It is very odd that the theory of gravity, originated by Newton and completed by Einstein, should stand now in majestic isolation, a Taj Mahal of science, having little if anything to do with the rapid developments in other branches of physics.

That isolation is ended. In the bustling city of present-day physics the Temple of Gravity is engulfed by modern industries: here the smoking chimneys of Relativistic Astrophysics Inc.; there the lathes of the Penrose Twistor Company; over there, the ovens of Hawking's Famous Gravitational Exploders; while all around the builders are striving to encase the entire conurbation in the Dome of Supergravity. Indeed admirers of the splendid symmetry of the einsteinian Taj Mahal may regret the disappearance of the pristine landscape.

Brilliant young physicists are carrying on where Einstein left off, stretching and testing his ideas to breaking point, at the limits of imagination and technique. Usurpation may be near at hand. As the reader will be able to see for himself, it now looks very unlikely that Einstein's ideas can be falsified in any superficial sense. Experiment after experiment bears them out. Occasional exceptions crop up but they tend to crumble away: apparent facts are often wrong, good theories seldom so. The successor to Newton and Einstein must instead look for extreme conditions which prove the equations inadequate and probe into the half-hidden electronics of the universe, short-circuiting some of its basic components. And once again it may seem that the universe quakes, as it did to Einstein's contemporaries.

So do not be deterred from trying to understand what Einstein said, by any thought that he may become outdated. On the contrary, if you have not yet felt the ground move under your feet while contemplating his ideas, you have missed the frisson of the century. The ideas are not entombed in textbooks that students mug up for their examinations. In so far as present-day scientists understand the universe we live in, Einstein's theories are the bedrock. He supplied no finished work of art to be admired but not touched by his successors. He created an imaginative framework for them, and specific theories to serve, directly or indirectly, in almost every serious attempt to consolidate and extend his understanding. It *is* Einstein's universe.

A recent and spectacular example of Einstein's ideas in action is the explanation they offer for a stupendous whirlwind of stars discovered in the galaxy M87. 'M' stands for Charles Messier who hunted comets from a tower in Paris 200 years ago. To avoid wasting time on fuzzy objects in the sky that were *not* comets, he decided to map them, to be rid of them as it were. In 1784 Messier published a catalogue of about a hundred of these annoying smudges. They are now known to include the most fascinating objects in the sky. Quite unaware that he would be remembered for his catalogue more than for his comets, Messier resumed his main work as an industrious inhabitant of Isaac Newton's universe, in which comets were the most amazing phenomena.

At Kitt Peak in Arizona in the late 1970s a group of astronomers used the 150-inch telescope to peer into the heart of M87, one of Messier's discarded objects lying in the constellation of Virgo. From London, Alec Boksenberg and Keith Shortridge had brought special electronic equipment capable of counting individual particles of light gathered from the depths of space. Another Englishman, Wallace Sargent, had joined them from the California Institute of Technology. Roger Lynds, an American astronomer based at Kitt Peak, and David Hartwick from Canada completed the observing team. They were all inhabitants of Albert Einstein's universe, wherein the prodigies were not comets but black holes.

When Einstein promulgated his law of gravity in 1915, the possibility of nature creating black holes—dark pitfalls in space from which nothing could ever escape—was implicit in the equations. And in April 1978 the Kitt Peak team announced evidence for the existence of a giant black hole in the heart of M87. It was a shocking thing, billions of times more massive than the Sun and capable of gobbling entire stars or clusters of stars, with ease.

As a galaxy, M87 is a vast agglomeration of stars like the Milky Way, the discus-like galaxy of which the Sun is a modest member among hundreds of billions of other stars. But M87 is even bigger, and more ball-like. It is a conspicuous member of a collection of galaxies known as the Virgo Cluster, which sprawls across a large area of the sky comparatively close to us. By present estimates M87 is about 50 million light-years away—that is to say, the light entering a telescope tonight, travelling at 186,000 miles a second, had to set off on its journey from M87 more than 50 million years ago, when our ancestors were primitive tree-dwelling primates. The astronomers' use of the travel-time of light to judge distances was ready-made for the marriage of space and time at which Einstein officiated. And his cosmic equations of 1917 implied that the whole universe could be expanding, with galaxies moving apart at high speed. That turned out to be the case and M87 is travelling away from us at a speed of about 700 miles per second.

In a minority of galaxies, astronomers discern great upheavals. M87 is one of the closest of these special objects. When radio astronomy began in earnest after the second World War, M87 was one of the first objects detected as a strong source of radio waves. Looking more intently, optical astronomers made out a faint luminous jet protruding from the galaxy like a lollipop stick a billion billion miles long, and blue in colour. Two decades later, when rockets and satellites going above the Earth's atmosphere detected X-rays coming from various directions in the sky, many of the X-ray sources were inside the Milky Way Galaxy; but M87 was one of the most conspicuous X-ray sources beyond it.

Altogether, M87 was very energetic. Yet it was a comparatively mild-mannered member of those special classes of galaxies which plainly radiated far more energy than they ought to—more energy, that is to say, than could be accounted for by the ordinary burning of all the stars of which they were composed. These exploding galaxies seemed to be related to the quasars which were discovered in the early 1960s: very small, distant objects, also giving off intense energy. What could be powering such violent eruptions far away in space? The theorists canvassed all sorts of ideas, from catastrophic collisions between large numbers of stars to the action of 'anti-matter', which is capable of annihilating ordinary matter. Einstein's theory of gravity came to the rescue and in the 1970s the opinion of many astronomers hardened in favour of black holes as the energy-source of exploding galaxies and quasars.

The idea was that a great star-swallower lay in the heart of a galaxy. When stars or gas came close to it they swirled in, faster and faster, like water approaching a plug-hole. The falling matter would radiate ever-more intense energy, up to the moment of oblivion. This process could also hurl out jets of matter from the whirlpool. The small-looking quasars were interpreted as very distant exploding galaxies, in which the central turmoil was intense enough to be plain even when the outlying stars were far too faint to be seen.

The team at Kitt Peak wanted to know how fast the stars in M87 were moving. The stars of any galaxy orbit around its centre, in much the same way as the Earth travels around the Sun; the Sun and we with it are orbiting around the centre of the Milky Way at 170 miles per second. If there is a massive collection of matter at the core of a galaxy, stars near to the core will move faster in their orbits. Discover the speed of the innermost stars of the galaxy and you can, in effect, 'weigh' the core.

That required a precision impossible before Boksenberg developed his electronic light-detector in 1973. It is called the image photon counting system; it registers individual particles of light, or photons, when they dislodge electrons from a sensitive surface exposed to the light. Einstein's interpretation of this 'photo-electric effect' established the modern view of the nature of light and the energy it carries; he was the discoverer of the photons that Boksenberg counted.

In M87 the astronomers looked for a smearing or de-tuning of the frequencies of particular kinds of light. Break up starlight into a spectrum, like a rainbow, red at one side and blue at the other; you will then see dark or bright lines at particular positions along the spectrum. They are like stations on a radio tuning dial and these lines correspond to light of precise frequencies absorbed or emitted by particular kinds of atoms in the stars. But for stars in rapid motion, like those swirling around the centre of M87, the frequency of the light changes. Peter Young at the California Institute of Technology used the results from Kitt Peak to build up a picture of the galaxy. The stars near the core of M87 were orbiting at about 250 miles per second. The mass of the core, necessary to sustain such motions, was 5000 million times heavier than the Sun. If there was an enormous number of stars at the core, corresponding to that mass, the heart of the galaxy would be very bright indeed. If, perhaps, dust were obscuring the view, you would expect the light to be reddened, like the Sun at sunset. But in the core of M87 the astronomers saw neither a dazzling mass of stars, nor a dusty red effulgence, but a lacklustre glow with a bluish tinge. In short, the appearance of the core contradicted the calculation of its mass unless it contained a black hole.

The Kitt Peak observations of M87 indicated that the central mass was confined within a region with a diameter of 700 light-years at the most—already a tight enough squeeze for the putative billions of stars. But if the black-hole theory is right the entire mass must in reality lie inside a sphere only about one light-day in diameter. Before the end of 1978, Boksenberg took his electronic detector to the 200-inch telescope on Palomar mountain in California, for an even closer look at M87, while Sargent also turned to a worldwide combination of radio telescopes, in Spain and California, to try to narrow down the region of the central mass. Thus did Einstein's theories continue to inspire and instruct the most modern research.

If you are puzzling out the origin of the energy in M87, or in any other violent object in the universe, you have a cosmic rule of thumb. The equivalence of mass and energy, as specified by Einstein, prescribes the maximum energy extractable from any aggregation of matter whatsoever. His formula $E = mc^2$, still amazing and frightening more than seventy years after he wrote it down, will be the subject of the next chapter.

2 : The Wasting Sun

High-speed motion changes the apparent energy of objects.
A moving luminous object seems to shed energy of motion.
Einstein inferred that light must be heavy.
Mass and energy are equivalent: $E = mc^2$.
Matter is frozen energy.

Relativity extends the human art of 'seeing the other fellow's point of view' into the realm of physics and astronomy. In his social behaviour, each person is conscious of how his actions look to other people, and large pieces of his brain seem to be involved in this task. He relies on this awareness of 'self' and 'others' to avoid traffic accidents and brawls. Similarly in the world of matter and energy you can ask, for example, how the Sun would appear to an astronomer in the vicinity of a distant star, and come quickly to the conclusion that it will itself look like an undistinguished star. More interesting effects arise if that alien astronomer is travelling at high speed towards the Sun: earthbound physicists reason that their white Sun will turn blue, from his point of view.

A difference exists, though, between the social world and the realm of inanimate matter. Human beings and many other animals adjust their appearance and actions according to who is looking at them. Inanimate matter does not do so: it goes on behaving as before regardless of whose telescope is trained upon it. Provided that the observation does not involve any significant interference, you do not expect to change the physical world just by looking at it. But the way it appears to you can certainly change.

If an astronaut flies at high speed past the Earth he is at rest, from his point of view, with the Earth hurtling past him. He therefore judges the Earth to have enormous energy of motion. Of this his colleagues on Earth are insensible. Therefore it has to be simultaneously correct to say that the Earth has great energy of motion and no energy of motion; the astronaut's point of view is just as valid as the view of learned men confined to the Earth.

When descriptions of the world disagree, the risk arises that the laws of physics specifying the behaviour of energy and matter may seem to be different to people travelling at various speeds. The astronaut might start predicting the Earth's behaviour differently. But no such rumination about appearances can alter the Earth's behaviour. It must continue to orbit steadily around the Sun, whatever the opinions of astronauts: the ideas of physics may falter, not the real world. But it should be possible to write down the laws of physics in such a way that their predictions remain correct regardless of the motions of the physicist. Not only is everyone's point of view equally valid, but all should agree on the essential features behind the appearances.

That was Albert Einstein's project. To accomplish it he was forced to do great violence to common-sense notions about time, as later chapters will reveal. But one of his greatest discoveries, the equation $E = mc^2$, involved such mind-troubling notions only obliquely. To follow him there we need little more than Doppler's effect, which changes the colour of light.

To sail and be unaware of the ship's forward motion was, for Lucretius and Newton, evidence of 'relativity' antedating Einstein. But stronger ideas about the curious effects of travel developed with the advent of railway trains. Passengers on the iron road immediately noticed that, even at a breathtaking sixty miles an hour when they were well aware of the joggling on the slightly uneven track, they could not *feel* any forward motion once the train had settled down to a steady speed. Rail travellers had the experience in a station of being momentarily uncertain about whether their own train was moving, or the one on the next track. Here were object lessons in the principles of relativity—that steady motion in a straight line is indistinguishable from being at rest, and that when two objects are passing each other at a steady speed you can equally well say that A is moving past B or B is moving past A.

More subtly, when people listened to the roar of a passing train or to its whistle, the pitch changed. It went 'hee-haw' like a donkey: 'hee' during the approach and 'haw' when the source of sound was moving away. Most of us, even if we noticed the effect, would shrug it off as no more significant than the rush of the wind and the rattle of the rails. But not Christian Doppler, a Viennese physicist working in the early nineteenth century. He won immortality by describing the effect carefully, extending it from sound to light, and pointing out that the change in pitch or frequency could be used to measure the speed of an object approaching or moving away from you—even the speed of a distant star.

That was in 1842. Nowadays we use 'doppler radar' in speed traps and burglar detectors; we measure temperatures by the 'doppler broadening' of light due to the violent thermal motions of the atoms that emit it; and we know that the distant galaxies and quasars are rushing away from us because the doppler effect alters the frequency of their light. In short, although Doppler died a quarter of a century before Einstein was born he found an unbeatable way of measuring relative motion.

The reason for the doppler 'hee-haw' is simple enough. An onlooker hears a train's whistle because pressure waves of a certain frequency travel at a steady speed from the whistle to his ear. If the train is moving towards him, the speed of sound through the air does not change, but each successive wave has less far to travel than the one before, to reach his ear. As a result the waves are crowded together and arrive at a higher rate per second than they would do if the train were at rest. The frequency of the sound is increased, which raises it in pitch. When the train is going away successive waves have farther to travel; they are spaced out and arrive with lower frequency or lower pitch.

The same effect occurs with light. Just as sound of different pitches possesses different frequencies, so does light of different colours. Light travels much faster than sound, and by comparison the fastest train is virtually at rest, so the onlooker would not see the train change colour as it passed him unless he had extremely sensitive instruments. But at higher speeds—among the stars for instance—the colour changes are easier to spot.

When white light goes through a prism it breaks into a spectrum of colours: red, orange, yellow, green and blue, shading off into violet. The frequency of blue light is about twice as high as the frequency of red light. As a swift luminous object approaches you the frequency of its light appears enhanced—it becomes 'blue' or at least shifts in the direction of the high-frequency, blue end of the spectrum. Physicists and astronomers call it a blueshift. Conversely, an object going away appears 'red'—its light is redshifted. I put 'blue' and 'red' in quotation marks because the colour change is not perceptible to the eye unless the relative speed of the source of light and the onlooker is very great.

Colour shifts in light and the equivalent changes of frequency of other forms of radiation, due to the doppler effect or gravity, crop up quite often in this book. In everyday psychological parlance, red is regarded as a 'warm' colour and blue as 'cool'; for a physicist the sense is the opposite. Blue, the high-frequency light, corresponds with high energy and high temperatures, while red, the low-frequency, represents lesser energy and cooler conditions. Saying that an object is 'red hot' implies a lower temperature than if it is 'blue hot'. More generally, a 'redshift' is scientific shorthand for the reduction in frequency and energy of any form of light, including all the invisible forms; a 'blueshift' is an upward change in frequency.

The precise reckoning of the doppler effect was a matter of great importance to Einstein, and he found that light did not behave in exactly the same way as sound. Because sound waves travel through a medium—the air—the extent of the doppler shift depends on whether the source of sound is moving towards the listener or the listener is moving towards the source of sound. Believing, wrongly, that light waves travelled through an 'aether' in space, Doppler's nineteenth-century successors supposed that the doppler shift for light also depended upon whether the source of light—a star, say—was moving towards the onlooker, or the onlooker was moving towards the star.

In Einstein's democratic universe, that cannot make any difference: all that matters is the relative speed of the star and the onlooker. The correction is a small but critical one—critical because blue light is inherently more energetic than red light, so that changing the colour of light affects its energy, and the universe runs on energy.

The concept of energy in physics is not very different from its meaning in everyday life. An 'energetic' person works and plays hard, and makes things happen around him. Nations go through 'energy crises' when they wonder how to keep their machinery turning and their buildings warm. For the physicist energy means the ability to produce changes in the world: to make objects move (energy of motion), to cause sunburn (radiant energy), to make plants grow (chemical energy), to set atoms in random agitation (heat energy), and so on. The greater the energy, the more spectacular are its possible effects. One form of energy can be converted into another. For example, in a car engine the chemical energy of the fuel changes into heat energy, which increases the pressure of the gases in the cylinders and so drives the pistons, creating energy of motion. Nature keeps strict accounts of energy and the total energy in the universe never changes; you can only shuffle it about.

Einstein knew all this, and he realised that his democratic correction to the doppler effect on light, which made a subtle alteration to the frequency and hence to the energy of light seen during relative motion, threatened to throw the universe out of joint. In preventing it from doing so he granted the universe a reservoir of energy surpassing the wildest imaginings of his contemporaries.

The formula for calculating the redshifts and blueshifts correctly, which Einstein wrote down, was not particularly difficult, but you can avoid the mathematics entirely by going to extremes. If a star were travelling away from you at the speed of light, its light would be redshifted completely. The waves of light would be stretched flat and their frequency would be zero; their energy content would also be zero. If, on the other hand, the star were travelling towards you at the speed of light, its light would be blueshifted completely. The waves of light would be heaped on top of one another; their frequency and energy-content would be infinite. You would be vaporised by intense rays one infinitesimal moment before the star, travelling amidst its own light, swallowed you up.

The doppler effect is not symmetrical in respect of energy. If a star goes away from you at the speed of light, what do you lose? One speck of starlight among countless billions in the universe. But if it comes towards you at the speed of light it is infinitely bright—far brighter than all those billions of stars put together. The energy gained in the blue-shift is greater than the energy lost in the redshift—much greater in the case just described.

Even at slower speeds the effect is still there. For example, at one-tenth of the speed of light the energy of light of the approaching star is enhanced by 10.55 per cent while the light of the star receding has lost 9.55 per cent of its energy. There is an average gain of half of one per cent in the energy emitted by the star, as judged by someone who sees it hurrying past him at one-tenth of the speed of light—or who himself travels past the star at that speed.

Now this is a curious state of affairs, because it means that the energy emitted by a star (or any other source of light) depends on who is looking at it, and on how fast he is moving in relation to the star. It leads us directly to what most physicists, including Einstein himself, came to regard as the most important single result of relativity theory. It is the idea that mass and energy are equivalent.

Einstein set out his own initial reasoning about the equivalence of mass and energy in a very short scientific paper published in *Annalen der Physik,* 1905, and entitled (in translation) 'Does the inertia of a body depend upon its energy-content?' It was a postscript to his first main article on relativity. He imagined what a source of light would look like to someone travelling at high speed. Adapting Einstein's argument to speak of the Sun rather than a 'body', and stripping it of its mathematics, I can paraphrase it like this.

Seen from a normal vantage point, the Sun gives out equal quantities of light in all directions. Suppose that an astronaut is dashing past the Sun at high speed. He will see the light altered in frequency by the doppler effect. As he approaches the Sun on the near side it will seem 'blue' and, going away from it on the far side, he will see it 'red'. But the discrepancy in energy sets in, as already noted. If the astronaut estimates the light energy coming off the Sun, by averaging the energy of the blueshifted and redshifted light, it is greater than he would record if he were stationary near the Sun and seeing it in its 'true colours', unaffected by the doppler effect. In the high-speed astronaut's judgment the Sun is giving off light energy at a greater rate than astronomers on the Earth, for example, would estimate. But his point of view is, according to Einstein, just as legitimate.

If the Sun is shedding energy faster where does the extra energy come from? The only possible source is the energy of motion that the Sun has by virtue of the fact that, relative to the astronaut, it is rushing past at high speed. The simplest way in which a moving object can shed energy is by slowing down. But the astronaut's speed is steady, so the Sun cannot change speed in miles per second relative to the astronaut without moving from its normal position. It would be a disorderly universe indeed if we could move the stars about just by looking at them.

There is, though, another way in which a moving object can reduce its energy of motion, and that is by reducing its mass. The Sun's energy of motion depends on its mass in tons as well as on its apparent speed relative to the astronaut. Therefore, by Einstein's reasoning, the Sun must be losing mass. Recapitulating the argument so far: the relative motion of the Sun, as seen by the high-speed astronaut, increases its apparent output of light energy; that extra light must be supplied from the Sun's apparent energy of motion; the apparent speed of the Sun cannot change, so its mass must change instead. In short the Sun has to lose mass in order to give off extra light and satisfy the astronaut's view of events.

It is a short step from that to saying that *all the light* given off by the Sun reduces its mass, even when seen by an astronaut who slows down to rest in relation to the Sun. For Einstein this step required a very simple piece of mathematics. For us, sufficient reason is that nature cannot distinguish between 'light' and 'extra light'; if radiating one involves a loss of mass, so does radiating the other. To keep the accounts of energy straight, the Sun must lose a certain mass in tons, in proportion to the energy of light that it radiates. In fact, measuring the Sun's brilliance, astronomers calculate from Einstein's theory that our mother star is losing mass at a rate of four million tons a second. But that mass is not disappearing from the universe; the light which the Sun is throwing out in all directions is itself heavy.

Then comes Einstein's master-stroke. Strictly speaking, all that his argument shows is that light has a certain mass associated with it—an important but not altogether surprising deduction. Yet light was formerly regarded as pure energy, and Einstein considered that the same argument would apply to all other forms of energy—not just the rays akin to light, but energy of motion, heat, chemical energy and so on. Furthermore, he saw intuitively that, if energy possessed mass, mass itself must be regarded as a form of energy. In that case ordinary matter was frozen energy. He wrote in 1905:

The fact that the energy withdrawn from the body becomes energy of radiation evidently makes no difference, so that we are led to the more general conclusion that the mass of a body is a measure of its energy-content. . . . It is not impossible that with bodies whose energy-content is variable to a high degree (e.g. with radium salts) the theory may be put successfully to the test.

The aside about radium salts referred to the then-recent discovery of radioactivity, in which ordinary-looking materials throw out remarkable amounts of energy. In the old physics there was no explanation for the source of that energy. Einstein pointed to the source. If ordinary materials are frozen energy they have, in principle at least, enormous inner reserves on which to draw. Only a small part of that energy need be released, by some internal rearrangement, to permit the phenomenon of the 'radium salts'.

Einstein's formula $E = mc^2$ expresses the equivalence of mass and energy. In it E is energy, m is mass and c^2 is the square of the speed of light. The c^2 comes in only because of the conventional ways in which physicists reckon energy and mass. You could just as well, and more simply, write $E = m$ and adjust your units of measurement to suit. But the c^2 is not quite *de trop:* it is a very large quantity by conventional standards and it tells us that even a small mass represents a very great deal of energy indeed. The little formula sums up all action and creation in the universe.

3 : Energy of Creation

All matter possesses enormous 'rest-energy'.
The Sun releases a little of the rest-energy of matter.
Nuclear bombs and reactors liberate similar amounts on the Earth.
Particle accelerators create new matter routinely.
The rest-energy is the energy required to create matter.

Direct and accurate confirmation of Albert Einstein's intuition about energy did not come quickly. He announced the equivalence of mass and energy when he was twenty-six years old, in 1905, several years before Ernest Rutherford's discovery of the small, heavy nucleus that lies at the heart of every atom. Thereafter the physicists had to invent methods of weighing individual atoms very precisely, and develop the first 'atom-smashing' machines, or particle accelerators. In 1932 another young man, John Cockcroft, was to be seen skipping down King's Parade in Cambridge exclaiming to his acquaintances: 'We've split the atom. We've split the atom!'

Cockcroft and Ernest Walton built an electrical accelerator for atomic particles and used it to shoot protons, the nuclei of hydrogen atoms, at a target of lithium metal. On hitting the nucleus of a lithium atom, a proton would merge with it and cause it to split into two new nuclei—of helium, in fact. The masses of two helium nuclei, added together, are slightly less than the combined masses of the hydrogen nucleus (proton) and the lithium nucleus, which made them. So matter disappeared in this nuclear reaction.

The experimenters could 'see' the pairs of helium nuclei that flew out from the target in opposite directions. Each flying nucleus caused a small flash of light when it hit a fluorescent screen, like a miniature TV screen. Cockcroft and Walton estimated the energy of motion of the helium nuclei by seeing how far they could travel through the air. To within an experimental accuracy of just a few per cent, the energy of motion of the helium nuclei gauged by Cockcroft and Walton fully accounted for the missing matter, in accordance with Einstein's formula $E = mc^2$.

You could say (and many people do) that a certain amount of mass has been abolished and converted into the energy of motion of the helium fragments. But that is a rather sloppy way of talking about events in

Einstein's universe. Mass and energy are not just interconvertible, in the way that I can convert dollars into gold or vice versa. They are the same thing—mass-energy. In the Cambridge atom-splitting experiment the flying helium fragments together possess, for a moment at least, exactly the same mass as the combined mass of the particles that produced them. Their mass-energy of motion is added to their conventional mass. They lose it as they slow down, handing over their energy of motion to other atoms in their neighbourhood, which share out their 'excess' mass as well; the energy is converted into heat. The mass-energy then remaining with the helium nuclei is often called the 'rest mass', and it corresponds to the traditional idea of mass—an amount of matter. But as matter is just a particular, frozen form of mass-energy, I prefer the term 'rest-energy' for conventional mass. Every other form of energy has mass associated with it. For example, a pressing iron is slightly heavier when it is hot than when it is cold, and a car has more mass when in motion than when at rest.

Nature was able to keep all of this secret until Einstein came along, even though his predecessors were not stupid, or careless in their measurements. The rest-energy of ordinary objects is enormous and, conversely, a large amount of energy possesses very little mass. All of the energy used by mankind in the course of a year has a mass of only a few tons. A hot iron is heavier by only one part in a million million of its weight when cold. The energy of motion of an *Apollo* spacecraft hurtling towards the Moon added to its mass only a few thousandths of a gram—which the engineers could safely ignore in calculating the rocket-power required to dispatch it.

For most motion with which human beings are concerned, the changes in mass are negligible. With accelerated sub-atomic particles, though, the energy of motion becomes substantial compared with the rest-energy. In a television receiver the beams of electrons that paint the picture on the screen are accelerated sufficiently to increase the mass of each electron by a few per cent. The world's most powerful accelerator of electrons is at Stanford in California and, from this great electric gun, two miles long, the electrons emerge from the 'muzzle' about 40,000 times heavier than when they started. All of that extra mass is energy of motion.

If you burn hydrogen gas in oxygen gas you create water, but you also produce a lot of light and heat. Energy, or mass, has been expelled from the material. So even before you make any measurements you can be sure, by Einstein's rule, that the water must be lighter in weight than the hydrogen and oxygen that made it. Conversely you have to inject energy, or add mass, to break water up into hydrogen and oxygen. But the discrepancies are only a few parts in a billion and nineteenth-century chemists understandably believed that matter was neither created nor destroyed in chemical processes. If the Sun burned its hydrogen fuel in an ordinary chemical reaction with oxygen, it would consume so much material in pouring out four million tons of light a second that it could sustain its output for only about a thousand years.

The Earth intercepts very little of the Sun's outpourings, yet about 160 tons of sunlight fall on the Earth every day. Green plants absorb the Sun's rays and use their energy to build carbohydrates out of water and carbon dioxide gas, and thus power all life on Earth. As you read these words your brain is running on electrical and chemical energy, which is translating the word-strewn images in your flicking eyes into perceptions and thoughts. The brainpower of the entire human population is less than a billionth of a gram a second.

The young Einstein's pronouncement on energy gives a deeper understanding of the existence of human life in the universe. The fundamental sources of energy are those associated with the cosmic forces: electromagnetism, manifested in light, in chemical reactions and in living processes; the sub-atomic forces responsible for nuclear reactions; and gravity. All of them are governed by $E = mc^2$.

Understanding the power of the Sun, on which all life depends, is one of the important scientific consequences of $E = mc^2$. The source of sunshine was an ancient puzzle which became acute in the nineteenth century. Charles Darwin and the nineteenth-century geologists realised that life on Earth was at least some hundreds of millions of years old, but eminent physicists like Hermann Helmholtz and Lord Kelvin could see no source of energy to keep the Sun burning for so long. They supposed that the heat of the Sun and the other stars came from gravity. It would set energy free when material from far and wide fell together, and settled tighter and tighter: it was the incandescent fall of a meteorite, writ large. From the prevailing rate of emission of energy Kelvin estimated the life of the Sun at not more than thirty million years.

Einstein's discovery of the enormous energy latent in matter and the nuclear physicists' detection of small discrepancies in the masses of the elements pointed the way to a satisfactory explanation of solar energy. It had to suit the age for the Sun and Earth inferred from studies of radioactivity—several billion years. The answer lay in the fact that hydrogen, the lightest and by far the commonest material in the universe and the Sun, could 'burn' in a nuclear fashion.

In the core of a newborn star, gravity creates the very high temperatures needed to fuse together the nuclei of hydrogen atoms—the protons. In a series of 'thermonuclear' reactions, four protons make a nucleus of helium, the next heaviest element after hydrogen. And, as in the Cockcroft-Walton experiment, the mass of the product (at rest) is slightly less than the combined masses of the pieces that went into it. The helium nucleus is less than the sum of its parts. To put that another way: just as gravity can release heat by 'binding' a meteorite to the Earth or gas into a star, so nuclear forces can bind the constituents of nuclear matter tighter together, releasing heat in the process.

The loss of matter in converting hydrogen into helium in the Sun is about seven tons for every thousand tons of hydrogen burned. That is a small fraction of the rest-energy of hydrogen, but it allows for the release of far more energy than ordinary chemistry would permit—or gravity in the scenario of Helmholtz and Kelvin. The possible lifespan of the Sun on this basis works out satisfactorily at about twice the present age of the Sun and the Earth.

The astrophysicists offer a bonus, in an account of how the chemical elements were made. Virtually all of the atoms in our bodies that are heavier than hydrogen were fashioned in stars that expired before the Sun and Earth came into being. The elements were scattered through space and then reassembled in the Solar System at its formation. When a star runs out of hydrogen in its hot core, it begins to burn helium to make carbon and oxygen. This process can continue through successively heavier elements, until silicon is burning and making iron. Up to iron, the process continues to abolish matter and release energy until, from every thousand tons of the primeval hydrogen, almost ten tons has been converted into heat and light. Iron is the most stable element and you cannot extract further energy from it by nuclear processes.

Cooking up the elements heavier than iron—gold or uranium, for instance—requires a net input of energy. It is a drain on the energy of the star and can be accomplished on a large scale only during the frenzy of nuclear activity that accompanies the explosion of a senile star. That is why gold and uranium are much scarcer than oxygen or iron. There too is the explanation of the fateful instability of heavy elements, which manifests itself in natural radioactivity and nuclear fission and prevents elements heavier than uranium from surviving. Because energy had to be *added* to their nuclei when they were being fashioned, they gained 'excess' mass by the interconvertibility of energy and matter.

Among the heaviest elements of all, the excess mass or rest-energy is sufficient to cause the nuclei to break up spontaneously. Typically uranium, radium and similar radioactive elements throw out small pieces—the most massive pieces being nuclei of helium. They continue doing that until they have shed enough rest-energy to become stable:

they change into lead or bismuth, the heaviest stable elements. The trickle of natural radioactive energy into the rocks of the Earth causes earthquakes and volcanoes.

Some nuclei of heavy elements can break up far more dramatically, especially when perturbed by a small input of additional energy. In particular uranium-235, a relatively rare constituent of uranium metal, will undergo fission—split into two large pieces—when struck by a low-energy neutron. A neutron is an electrically neutral sister of the proton and a major constituent of nuclear matter. During fission, other neutrons are let loose, which in turn can split other uranium-235 nuclei, in a chain reaction. Similar nuclear fission occurs in plutonium, a man-made metal formed by cooking the common uranium-238 with neutrons in a nuclear reactor.

With fission, nuclear energy becomes conspicuous on the Earth, because a significant part of the rest-energy of the uranium-235 or plutonium is suddenly set free. Yet only one ton in a thousand tons of the nuclear material disappears in the process. It is significant only because of the immense conversion rate for matter into other forms of energy. None of this technical detail was known to Einstein when he wrote $E = mc^2$. It is all the more astonishing that his ruminations about the appearance of a source of light when you are moving past it should have led him to conclusions that enabled astronomers to understand how stars work, and other scientists to compute the accessible energy latent in nuclear matter and then release it in bombs and nuclear reactors. The potency of $E = mc^2$ was demonstrated to the world with the A-bomb explosions of uranium at Hiroshima and plutonium at Nagasaki, in August 1945.

By the later invention of the H-bomb, mankind took the first grim step towards reproducing on the Earth the nuclear fusion by which the Sun and the stars burn. In the H-bomb, a uranium bomb serves to create very high temperatures similar to those prevailing at the heart of the Sun. Then lightweight elements will burn, augmenting the explosion both directly and by making the uranium explode more efficiently. The bomb-makers did not attempt to use the simple protons of ordinary hydrogen (they would react too slowly) but chose for their fusion explosive the heavy form of hydrogen, plus lithium. Other physicists began strenuous efforts to achieve 'controlled thermonuclear reactions' for a peaceful use of fusion power. Success in this project promises to solve the world's energy problems, if the H-bombs do not solve them in another fashion first. For better or worse, mankind's future is wholly bound up in $E = mc^2$.

After Einstein wrote down his formula no one quite knew what sense there was in saying that each ordinary piece of matter had enormous quantities of energy associated with it. Rest-energy is, to be sure, one of the 'absolutes' in relativity: everyone, regardless of his own movements, can agree on the rest-energy of an object, after discounting the energy of motion. But as a promise of energy it seems an empty declaration. Even the most vigorous nuclear reactions can tap, at most, one per cent of the rest-energy of the fuels. What about the other ninety-nine per cent?

The fact that rest-energy, the conventional mass of objects, cannot easily change into other forms of energy is reassuring. If human beings could explode with the energy of 1000-megaton H-bombs, in accordance with Einstein's equation, their social relations might become a little strained. But evidence that the rest-energy of ordinary objects was more than an abstraction came with the discovery of the creation of matter. Physicists working with atomic particles found that light of sufficient energy—very energetic 'gamma-rays' to be precise—could make fresh sub-atomic particles. The energy of light was transformed into matter. And the minimum energy needed to create a particle depended on its rest-energy, as specified by Einstein's equation.

The story was a little richer than that—as you might expect for the creation of matter in previously empty space. In practice, nature has to create two particles at once. The total number of durable particles in the universe cannot casually change, but that is avoided by the creation of an 'anti-particle' alongside the particle. For example, from gamma-rays with energy equivalent to *twice* the mass of the electron you can make an electron and an anti-electron. Anti-particles are very like the corresponding particles but their opposite—so exactly opposite that when a particle meets its anti-particle they annihilate one another and disappear from the universe. All that remains is a 'puff' of gamma-rays—a reversal of the process of creation.

The existence of anti-matter was predicted by a young English theorist, Paul Dirac, when he applied Einstein's relativity to the theory of electrons. Dirac's bizarre and rather spooky idea was vindicated in 1932 by the discovery of the first known anti-electron, or 'positron'. It was produced by the cosmic rays—particles that fall on the Earth from outer space. Eventually particle accelerators became powerful enough to create much heavier anti-particles: anti-protons, the anti-world's equivalent of the nucleus of the hydrogen atom. That was done by Emilio Segré and Owen Chamberlain at Berkeley, California, in 1955.

By the 1970s the production of matter and anti-matter had become entirely routine—confirming Einstein's formula in every pulse of the big accelerators. They create not only the familiar sub-atomic particles but heavy particles with peculiar behaviour designated by names like 'strangeness', 'charm' and 'beauty'. They reveal fundamental rules operating in nature's organisation of matter. And $E = mc^2$ is no longer just the secret of the Sun's energy and the nuclear bomb—it is, quite literally, the law of all creation.

So now the deep meaning of the rest-energy of matter becomes apparent. It is the energy needed to create matter in the first place. It could be released in full by annihilation in encounters with the appropriate anti-matter. If God were to create Adam from scratch, he would need to marshal all the energy of a 1000-megaton explosion. Indeed, he would require twice as much, in order to create anti-Adam at the same time.

To anticipate what I shall be saying later about the history of the universe: everything probably began in a Big Bang. Enormous numbers of particles and anti-particles formed and disappeared in a frenzy of creation and annihilation. As the universe cooled and the energy of the prevailing radiation grew less, annihilation supervened, until only a billionth part of all that matter remained. A slight imbalance in particle numbers eliminated the anti-matter, leaving a 'small' residue which constitutes the matter of the present universe.

The atoms of our bodies have been reworked and recombined many times since then, but the basic ingredients—protons and electrons—were all fashioned in that unimaginable furnace. Had you existed then, observing the microphysics of creation, there would be no doubt in your mind about the meaning of rest-energy. You would have seen rays converted into matter, and vice versa, at the high rate of exchange specified by Einstein. Nowadays the universe is far cooler, the creation of new matter is a rare event and the fortunate dearth of anti-matter allows existing matter to conceal the energy of its creation.

The kinship and difference between matter and light-like rays appear in the furnace of creation. The rays have energy equivalent to mass which, in the Big Bang, was typically comparable with the masses of the particles of matter. The mass-energy of the universe did not change one jot during the interconversion of rays and particles. But light has no rest-energy. It can be absorbed and converted into another form of energy, but light cannot be brought to rest. At its birth, it accelerates infinitely rapidly and starts rushing along at the highest possible speed; that is the nature of light. The nature of matter, on the other hand, is to travel more slowly and to change speed sluggishly—that is what we mean by conventional mass or rest-energy.

4 : The Ultimate Waterfall

An object falling downwards loses a little rest-energy.
Gravity can arrest light and create a black hole.
An object falling into a black hole releases a lot of rest-energy.
Collapsing stars create intense gravity.
Rest-energy released by gravity can explain violent cosmic events.

Just as the matter of your body is energy waiting to be liberated should you ever encounter anti-matter, so you can think of it as energy which might also be released, in part or entirely, if you should be unfortunate enough to fall into a black hole. To look at the implications of black holes, as knots of gravity that bind matter with unbreakable bonds, is another way of grasping the meaning of $E = mc^2$.

The idea of extracting a little energy from objects falling under gravity is familiar enough. For instance, nations well endowed with mountains generate a good deal of electricity from waterfalls. River water rushing down a mountainside drives the turbines of a hydro-electric power station. The greater the 'head' of water, from the start of its fall to the turbines, the more energy you can win from it. The water of most rivers does not drop any farther than the sea-level. But you could run water into a deep depression, as in the Dead Sea or the Qattara Depression in the Middle East, and so gain more energy from it.

What is the limit to that process? Imagine a mad engineer who is dissatisfied with the 'head' of water that he is given to work with. He proposes to perfect his hydro-electric scheme by digging a very deep tube well in the ground, right down to the centre of the Earth. In principle the engineer could gain a lot by dropping his water all that way, but not as much as he might expect. As he digs deeper, the rocks above begin to neutralise the gravity of the rocks and the iron core beneath. At the centre of the Earth, gravity is zero. That restricts the amount of energy he can obtain by this strategy.

In his cosmic madness, our engineer will see what he has to do: compress the rocks of the Earth into a very small volume, while keeping his supply of water in orbit, ready to feed the ultimate waterfall. Let the practical difficulties speak for themselves; we talk of cosmic principles. If he devises suitable hoops for squeezing the Earth from a diameter of 8000 miles to less than an inch, its gravity will be sustained for a much longer drop, and it will become extremely strong in the vicinity of the miniaturised Earth. In fact the engineer has created a black hole, where the grip of gravity becomes so great that even light cannot escape from it. Now if he drops his water in, all the engineer's efforts will be rewarded: each drop of water will accelerate to almost the speed of light. Recovering most or all of the rest-energy is now a possibility, and every drop of water becomes equivalent to a hundred tons of high-explosive.

For the best results our engineer should consult Roger Penrose, one of the most distinguished of present-day relativists. In the 1960s he figured out many of the basic features of black holes. He also envisaged a machine that would enable an advanced civilisation to meet all its energy needs by dropping its garbage into a black hole. A prime requirement is that the black hole should be spinning.

A spinning black hole has a fringe zone around it, where objects falling into it retain a chance of escaping. In Penrose's machine, the people drop buckets containing their garbage into the black hole. They might quickly spiral into oblivion but, before that happens, each bucket dumps its load. In doing so, it gains energy of motion equivalent to the entire rest-energy of the garbage (or even more, but that is a complicated story). The energy is ample to carry the bucket out of the grip of the black hole and back to base. There the returning buckets engage with a flywheel which converts their energy of motion into electricity and also slows down the buckets so that they can be used again.

Water would do as well as garbage. Or, remembering the rest-energy of the human body, the inhabitants of Black Hole City might elect to inter the mortal remains of their loved ones in the black hole. The bucket becomes a recoverable coffin and Auntie Lou's bequest to her relatives is a million million kilowatt-hours of useful energy, which is worth billions of dollars at present terrestrial prices. These industrial and necrophorous technologies are fictional but the rest-energy of ordinary matter is not. And nature evidently knows of ways of extracting a lot of it by processes in which matter falls into black holes or on to other collapsed stars.

The nineteenth-century physicists were wrong about the energy and lifespan of the Sun yet more right than they knew about gravity as a source of cosmic energy. By compressing matter, gravity can release a substantial part of its rest-energy. The connection between $E = mc^2$ and gravity leads us to the fate of the Sun, and to exploding stars and galaxies.

The Sun and most of the visible stars in the sky are in mid-career, steadily burning their hydrogen fuel. The nuclear heat generated in the hot core pushes outwards on the material of the star, opposing the strong gravity which tends to pull the stuff of the star together. The contrary forces reach a compromise in which the Sun, for example, has an average density of matter not much greater than that of water on Earth. At the core of the Sun the density rises to a hundred times the density of water.

When a star like the Sun begins to run out of nuclear fuel it will expand and flash (for reasons that need not detain us) and then collapse under gravity into a 'white dwarf' star. Such white dwarfs are known to astronomers: the first to be discovered was the faint companion of the most conspicuous star in the sky—Sirius. In a white dwarf, gravity compresses matter until the average density is a million times greater than water's. A white dwarf containing about the same mass of material as the Sun is only about as big as the Earth. In these circumstances gravity becomes very strong and the falling-together of matter, which creates the white dwarf, liberates substantial amounts of energy. In this last spasm of the dying star, the energy released in the collapse is equivalent only to a few ten-thousandths of the rest-energy of the matter of the star, considerably less than the energy produced by the nuclear burning of hydrogen during its earlier life.

Stars that are a good deal more massive than the Sun burn much more fiercely, have much shorter lives and end their days more spectacularly, in a supernova explosion. The star gives out more energy in a few days than in all the millions of years of its previous existence. The most celebrated explosion of that kind was recorded in AD 1054, by Chinese astronomers. The remains of the star in question can be seen still hurtling out from the explosion. They make up the Crab Nebula or M1, discovered by Charles Messier two centuries ago. At the centre of the explosion there remains, not a white dwarf, but a pulsar flashing regularly thirty times a second and broadcasting radio and X-ray energy as well as visible light.

A pulsar is a 'neutron star', in which the gravity has compressed the residual mass of the star into a ball only about ten miles in diameter. The density is extremely high and a thimbleful of matter in a pulsar would weigh 100 million tons. The other name for it, 'neutron star', refers to the peculiar state of matter that exists under such intense pressure. Gravity overwhelms the electric force, which gives atoms their normal size, and squeezes the nuclei of all the atoms together.

The nature and behaviour of pulsars is a fascinating theme in its own right but the salient point is that the collapse which makes a pulsar releases about ten per cent of the entire rest-energy of the stellar material. Here gravity far surpasses any nuclear reactions and begins to release a significant fraction of the energy that could be obtained by annihilating the matter completely.

In some cases, matter from a companion star rains on to the surface of the pulsar. The violence of the fall makes this material incandescent —not just red-hot or white-hot but 'X-ray-hot'. The X-rays pulsate regularly, like the radio waves and light which the collapsed star emits. The reason is that the pulsar has very powerful magnetic poles which guide the falling matter and create hot-spots on its surface. Because the pulsar is spinning rapidly, these hot-spots come into view and disappear repeatedly.

Some other X-ray stars, the most conspicuous being Cygnus X-1, show no such regular pulsations, although their emissions fluctuate violently. They are suspected black holes. In the case of Cygnus X-1, matter from a visible companion star, designated HD 226868, seems to be falling into the black hole. The black hole may have a mass six times greater than the Sun's. In the explosion of a giant star at the end of its normal life, the combined effects of gravity and an implosive reaction to the blast-off of the star's outer layers can create a black hole from the inner material of the star. In theory, and almost certainly in practice, they can overwhelm even the nuclear forces that give sub-atomic particles their normal size, and squeeze the inner material of the star to the extraordinary degree contemplated for the Earth by our mad engineer. The name 'black hole' was coined by John Wheeler, the doyen of American relativists and a man with a vivid turn of phrase. If you were fairly near to a star that had suffered extreme collapse you would not see it but the stars behind it would be blotted out, making a black hole in the sky.

From Einstein's theory of gravity, a black hole six times as massive as the Sun is reckoned to have a diameter of only about 22 miles. The black hole itself is not visible, but such a totally collapsed star creates the ultimate waterfall. Even without an artificial contrivance like Penrose's machine, matter swirling into a rotating black hole can release up to 40 per cent of its rest-energy, in the form of X-rays and other radiation. When a black hole is continuously fed with matter, its surroundings can glow extremely brightly. The falling matter exudes the energy as if in a dying shriek, before it disappears for ever.

Extremely massive black holes are possible in theory. They may have been produced either in the Big Bang at the birth of the universe or by the later aggregation of matter equivalent to very many stars. The suspected black hole in the nucleus of galaxy M87 possesses, as mentioned earlier, a mass estimated at five billion Suns. The swirling of stars and gas towards a plug-hole of that kind can expel, besides the radiation, jets of hot matter such as the one observed in M87.

Astronomers needed such a source of cosmic energy. The extraordinary intensity of the quasars shocked the people concerned in their discovery in the early 1960s. Radio astronomers were perturbed by the enormous amounts of energy coming from extended radio galaxies and from the very small, starlike or quasi-stellar radio sources, or quasars. But the full ferocity of quasars did not become clear until 1963, when radio astronomers in Australia pinpointed one of them and Maarten Schmidt, a Dutch optical astronomer working at Palomar, examined the corresponding visible object and deduced that the quasar was a long way away. The outpouring of radio energy and light from a very small volume of space was therefore enormous.

Schmidt told me: 'That night, I went home in a state of disbelief. I said to my wife, "It's horrible, something incredible happened today".' The universe seemed out of joint. Confronted with such unaccountable energy, some experts began speculating about mysterious new forces in nature. In the upshot, there is no need for a new force to do the trick. Einstein's gravity, in the extreme conditions of the giant black hole, wrenches rest-energy from falling stars and explains the quasars to the satisfaction of most astronomers. In this interpretation quasars are the small cores of violently exploding galaxies, far outshining the ordinary stars.

Conceivably many or all galaxies, including our own Milky Way, possess massive black holes in their cores. In that case, quasars and other intense outpourings of energy represent relatively short phases in the lives of ordinary galaxies. Astronomers imagine a black hole in the core of a galaxy swallowing gas and stars from its surroundings for perhaps fifty million years. Eventually it becomes starved of fuel because it has swept up all the vulnerable material, like a vacuum cleaner, and the surviving stars are orbiting at a safe distance.

The Sun and the Earth lie in the suburbs of the Milky Way and a quasar at the core of our Galaxy would appear like a persistent glare in the constellation of Sagittarius, bright blue in colour and as luminous as the Full Moon. It is not certain that the Milky Way was ever as spectacular as that: observations of the motions of the stars put an upper limit to the mass of any black hole at the galactic centre, of about five million times the mass of the Sun, or one thousandth of the object known as M87. Yet there *is* a compact source of radio energy in the centre of the Milky Way at present, which could be the supposed black hole grazing on meagre remnants of fuel.

A few astrophysicists, notably Philip Morrison and Kenneth Brecher of the Massachusetts Institute of Technology, remain thoroughly sceptical about black holes—which offend their sense of propriety in nature. While still allowing that gravity can crush matter and release the enormous amounts of energy pouring from X-ray stars or the cores of galaxies, they deny that the collapse or congregation of stars need proceed so far as to create a black hole. Spinning a massive star might stave off complete collapse for a long time, and new 'anti-gravity' effects may be discovered that prevent the eventual collapse into a black hole. But you can rule out black holes only by saying that Einstein's theory of gravity, enshrined in General Relativity, is wrong.

Conversely, proving the existence of black holes to everyone's satisfaction will be a necessary step in verifying General Relativity in extreme conditions. But many theories of gravity other than Einstein's would predict black holes behaving in roughly similar ways. Detailed observations of black holes in the future may enable astronomers to see whether they are einsteinian black holes or some other variety. For example, different theories give different predictions for the fastest rate at which the emissions from infalling material will fluctuate, and for influences of gravity and motion on the observed emissions.

Black holes in action provide a foretaste of Einstein's theory of gravity. When energy, primarily in the form of matter and its rest-energy, congregates together to make a massive object like a planet, a star or a black hole, it exerts an effect on other matter or energy in its vicinity. It sets up a gravitational influence, whereby energy interacts with energy on a huge scale. It is in the nature of the interaction to devalue the rest-energy of matter in an object nearing a source of strong gravity. Energy cannot be created or destroyed, so the missing part of the rest-energy reappears in other forms—the energy of motion of water rushing down a hillside, or the radiant X-ray energy of gas falling into a black hole.

The way in which the devaluation of the rest-energy occurs, in Einstein's theory, is by a disconcerting effect of gravity on time itself. Before pursuing that point in detail, I shall prepare the ground by digressing into Einstein's discoveries about the nature of light, which revealed a quite different sort of connection between energy and time.

5 : Einstein's Clock

Rearrangements in atoms produce light of precise energy.
Einstein discovered that light consists of particles.
The energy of light-particles fixes the rate of vibration.
Light can stimulate atoms to emit similar light.
Stimulated emission serves in laser beams and atomic clocks.

Knowing how atomic clocks work is not essential for understanding relativity. They did not exist when Albert Einstein developed his theories. But he himself discovered the two unexpected principles by which they run. In that sense Einstein was the grandparent of the atomic clock which became, half a century later, the basis of modern timekeeping.

Every atom in the universe is a natural timepiece because it absorbs and emits light at precisely defined frequencies. Most timekeeping relies upon counting a regular series of events: the alternations of day and night, the swings of a pendulum or a balance wheel, the vibrations of a quartz crystal, and so on. In the case of light, the vibrations are electric. The frequency of visible light is around five million billion vibrations per second. But there exists a great cosmic rainbow of light-like radiation. At one extreme, radio energy of extra low frequency vibrates only a few times a second; at the other end of the spectrum, gamma-rays possess frequencies a billion times greater than those of visible light. Yet all of these forms of energy travel at the same speed and have the same general character. When relativists speak of 'light' they usually mean electromagnetic rays in their broadest sense, and not just visible light.

For the light produced by a particular kind of event in atoms of a particular kind, the frequency is precisely defined. The 'event' is a rearrangement of pieces inside the atom which involves shedding or taking in energy in the form of light. Because the pieces of an atom can be arranged only in certain well-defined patterns, the energy of the light is equally well defined. And Einstein discovered that the frequency of light depends precisely upon its energy.

Einstein was a versatile physicist. He spent his early professional years as an examiner of inventions for the patent office in Bern in Switzerland. This work required him to exercise his mind on all manner of practical problems, but he had enough spare time to pursue his own thoughts and write theoretical papers. In 1905, shortly before he crystallised his first set of ideas about relativity and $E = mc^2$, he wrote the paper which was to win him his Nobel Prize while relativity was still controversial. In it, he declared that light consisted of particles.

Experimenters in Germany in the early 1900s were greatly puzzled by the 'photo-electric effect'. When light (ultraviolet rays in particular) shines on a metal surface, it knocks electrons out of it. A natural expectation would be that, if you weakened the beam of light, the individual electrons coming out of the metal would be less energetic. But the experiments showed that nothing of the sort happened. Fewer electrons emerged, to be sure, but those that did come out had individually just as much energy as before. For the physicists who supposed that light consisted simply of waves it was extraordinary, as if a gentle ocean wave rolled into a harbour, picked on a single ship and tossed it a hundred feet into the air. It made perfectly good sense, though, if light consisted of particles—each a bullet that would naturally concentrate its energy on a single electron.

The experiments had also shown that high-frequency light ejected the electrons from the metal surface with more energy than low-frequency light did. Einstein realised that the energy carried by a particle of light was simply proportional to its frequency, in cycles per second; double the frequency and you double the energy of the particle of light. This was the essential clue that the Danish atomic theorist Niels Bohr needed, in order to interpret the emission and absorption of light in terms of the rearrangements of the pieces of an atom.

All atoms of a given type, hydrogen for example, turn out to be perfectly identical: nature mass-produces them more faithfully than any man-made factory could do. This is the case not only on the Earth but throughout the universe. As a result, an astronomer can recognise particular events in particular kinds of atoms in a planet millions of miles away, or in a quasar a billion light-years away. When he spreads the light from the planet or the quasar into a spectrum of different frequencies he sees patterns of bright or dark lines, corresponding to emissions and absorptions in atoms. Each kind of atom generates a pattern of lines, like a fingerprint, and the astronomer can make inferences about the composition and general condition of the distant object. The frequency of any one line will not be exactly the same as it would be on the Earth. By the discrepancies he can detect the effects of magnetism and also tell (by Doppler's effect) how hot the object is and how fast it is moving relative to the Earth.

The high frequency of almost all kinds of light means that you can in principle subdivide each second of time into a large number of parts, and so measure very short intervals. But atoms have one overwhelming drawback as practical timepieces. Once an atom has rearranged itself, emitting or absorbing a single particle of light, it does not continue the process. It has had its little fling, at least until, at its leisure, the atom reverses the process and is ready to do the same thing again. As a result the light comes off at random from a collection of atoms. Each particle of light can be thought of as a brief train of waves, vibrating electromagnetically at the correct frequency but passing in a flash. In most natural circumstances, the particles of light come into being independently and the light is completely incoherent.

For keeping time you need an unbroken wave of light like the continuous waves of radio energy that engineers produce by making large numbers of electrons march repeatedly to and fro in unison, in an alternating electric current. To accomplish this you must create a steady supply of atoms, all in the same state and ready to undergo the same internal rearrangement. And then you must make each of them emit its particle of light at precisely the right moment, so that the individual wave-trains of light add up into one coherent wave, continuing indefinitely.

Einstein's second great theoretical discovery about the nature of atomic light was that coherence can be achieved rather easily. He announced it in 1916, when he had just completed General Relativity and the theory of gravity. He reasoned that an atom that is ready to rearrange itself in a particular way and give off light of a particular frequency can in fact be provoked into doing so by the arrival in its vicinity of another particle of light of the same frequency. In a phrase, stimulated emission of radiation occurs. And, like soldiers joining a marching column, the new particles of light fall precisely into step with the precursors. Radio astronomers have recently found this kind of process going on naturally in clouds of gas among the stars.

The word 'laser' comes from the initial letters of 'light amplification by stimulated emission of radiation'. Not until the early 1960s did physicists succeed in making lasers—lamps based on Einstein's principle which produce waves of light of extraordinary purity, coherence and intensity. Within twenty years of their first appearance, lasers were reading price tags in supermarkets, melting metal, and measuring the distance of the Moon. The novel image-making technique of 'holography' was a by-product. And for natural philosophers the laser completely reconciles the conflicting views of light, as particles or waves. Put the vibrant particles in step with one another and they make a regular wave.

The atomic clock came in a little earlier than the laser: it was a practical timekeeper by 1955, the year of Einstein's death. It depended on the same principle applied to very short radio waves, or microwaves.

The standard atomic clock uses a continuous beam of atoms of the element caesium, all ready to make a certain rearrangement involving the emission of microwaves. In a cavity inside the clock, the atoms stimulate one another to make the change and produce a continuous wave at a precise frequency. The wave is used to regulate the vibrations of a quartz crystal, which in turn can drive an electronic digital display showing the time of day. Caesium is not the only element used in atomic clockmaking: rubidium serves in many clocks for practical applications, and hydrogen in clocks for scientific work where exceptionally high precision is required.

Since 1967, officially and internationally, the reckoning of time is based on the caesium atomic clock. Nowadays one second is *defined* as 9,192,631,770 vibrations of the microwave radiation emitted by caesium-133 atoms during a specified atomic rearrangement. The readings from about eighty atomic clocks in government laboratories around the world are regularly pooled at the International Time Bureau in Paris. After some 'weighting' in favour of the steadiest clocks, the bureau notifies a democratic mean atomic time to the world. The atomic clock is far more reliable as a timekeeper than the rotation of the Earth, and the apparent motions of the Sun and the stars. Indeed geophysicists can study, with its aid, subtle changes in the length of the day. Every so often you may notice official time being put forward or back by a 'leap second', to correct for the discrepancies between atomic and astronomical time.

Some thousands of atomic clocks are in use around the world, so they are no longer a laboratory curiosity. They can withstand heat, cold and hammer blows. A typical commercial version is the size of a drawer and weighs about 55 lbs, although smaller ones are coming into service. Thanks to the atomic clock, time-measurement to an accuracy of a millionth of a second a day has become entirely routine; even greater precision for limited periods is possible with special atomic clocks for research purposes, which use hydrogen, not caesium. And in this real world of high-precision clocks the keepers of official time find themselves face to face with the consequences of relativity.

Einstein predicted that clocks would be affected by motion and by gravity, and sorting out these influences assumes a practical importance. For example, ships and aircraft using radio navigational aids such as Loran-C and Omega rely on precise measurements of the time of receipt of radio signals from various distant radio transmitters, to tell them where they are. These systems can be accurate to within a few feet, provided that the transmitters scattered around the world are carefully synchronised. But there is scope for errors if relativistic effects are not taken into account. And then the navigator himself, moving around, accumulates his own relativistic timekeeping errors. Unless he resynchronises his clock frequently at known positions, these errors grow to become significant. More generally, astronomers, geophysicists and official timekeepers in various countries would like to encase the world in a framework of synchronised time. They have artificial satellites to help them but the difficulties are not only technological but philosophical, because Einstein realised that there are inherent variations in timekeeping, from place to place and in accordance with motion.

In 1971 two American physicists, J. C. Hafele and Richard Keating, carried out the pioneering experiment of taking portable caesium clocks (four of them, for safety and reliability) right around the world in passenger jet aircraft. They compared them at the beginning and end of the journeys with the reference clocks at the US Naval Observatory in Washington DC. One circumnavigation was made eastwards and one westwards, with both journeys taking about three days. The result of the experiment was that the clocks no longer agreed about the time of day.

The eastbound clocks lost, on average, 59 nanoseconds (billionths of a second) compared with the clocks in Washington, while the westbound clocks gained 273 nanoseconds. In Newton's universe, there would be no accounting for the discrepancies in such highly reliable instruments. How could atoms disagree about the time? But the results were in satisfactorily close agreement with what Hafele and Keating expected in Einstein's universe. Taking account of the flight paths reported by the aircraft captains, they predicted from relativity theory a loss of 40 and a gain of 275 nanoseconds, for the eastbound and westbound clocks.

Two distinct effects on timekeeping were confirmed in the experiment. The first and more fundamental one is that clocks run faster at high altitudes, where gravity is slightly weaker. This affected both of the trips in high-flying aircraft in much the same way. The differences arose from a second and more subtle point in Einstein's theory, concerning the behaviour of clocks travelling in the same direction as the Earth's rotation, or against it.

This is a preliminary illustration of real relativistic effects on timekeeping; subsequent experiments have tested them with greater precision. As a word of encouragement for facing up to peculiarities about time, as they will unfold in later chapters, it is worth recalling *apropos* the circumnavigating clocks the grosser astonishment of the very first encirclers of our planet. When the survivors of Ferdinand Magellan's around-the-world expedition of 1519–22 arrived home they found that they had somehow 'lost' a day. The explanation might have been that they were all drunk one day—but in fact their log-keeping was meticulous. The Portuguese sages scratched their heads, and soon realised what had happened. If you record time by counting the sunrises, and if you go once around the world westward, you will see one less sunrise than your brother who stayed at home. Your solar clock runs slow.

Gaining and losing days was a shocking thing to our forefathers who supposed their days were 'numbered', with the dates of their impending deaths inscribed in some angelic ledger. Nowadays many travellers cross the 'international dateline' in mid-Pacific, where they lose a day going west or have the same day all over again going east. I admit that it still seems a little uncanny, but familiarity and a simple explanation softens the mystery. So let it be with relativity.

6 : Weightlessness

General Relativity deals with gravity.
A person falling freely feels no force of gravity.
In a freely falling spacecraft objects are weightless.
The Earth falls through the universe at a great rate.
Gravity creates tracks through space.

Albert Einstein reworked the theory of gravity in General Relativity and the confirmation of one of his predictions by astronomers in 1919 made him world-famous. He was acclaimed as a genius and became a symbol of all that was best in the strivings of the human intellect. Unfortunately relativity was also supposed to be incomprehensible—which Einstein himself tried to rectify. At that time a newspaper reporter asked him what had inspired his work on General Relativity. Einstein answered with a little parable: He had been triggered off, he replied, by seeing a man falling from a Berlin roof. The man had survived with little injury. Einstein had run from his house. The man said that he had not felt the effects of gravity—a pronouncement that had led to a new view of the universe.

Some people (including apparently Ronald W. Clark, whom I am here quoting) regard this story as a poor joke played to a foolish gallery. Although his tongue was in his cheek, Einstein's meaning was wholly serious and he expressed the essence of his theory in an exact and vivid form. Indeed, in an unpublished manuscript recently unearthed by Gerald Holton, Einstein wrote it down in terms that were only slightly more formal: 'There came to me the happiest thought of my life. . . . If one considers an observer in free fall, for example from the roof of a house, there exists for him during his fall no gravitational field—at least in his immediate vicinity.'

A man falling off a roof does *not* feel any force of gravity. In modern parlance, he is weightless. Falling needs no explanation, in Einstein's theory, because cosmically speaking it is the most natural thing that can happen to anyone. Only in *avoiding* falling does any force come into play. The weight you feel on the soles of your feet is pushing upwards, not downwards.

The reader may say, 'But the force of gravity is real—terribly real! It can drag an aircraft out of the sky. It can snatch a climber off a roof or a mountain face and smash him to pieces at the bottom.' Now, no one is denying the *effect* of gravity, sometimes terrible, often tiring, although generally useful in saving you from the inconvenience of falling off the planet. By all means call it a force if you want to—scientists certainly do so. The only question is whether that is the best way to describe it in order to understand its effects. A force acts on the climber before he loses his footing; a footing is a way of borrowing an upward force from the Earth to resist the effect of gravity. A force also acts on the climber when he hits the ground at the bottom—for a moment a much greater force than before because it has to absorb the speed of his fall. But in the course of his fall the only force acting on him is the inadequate resistance of the air. He sees the ground 'rushing up to meet him': relativistically that is an apt way of describing his predicament.

Human beings devote a lot of effort and brainpower to staying upright and safe against gravity in spite of the difficulties that arise from having only two legs. Much of their ability in this respect is innate. Let a newborn baby's head drop a little and he will clutch as if for his mother's hair, to save himself from falling. Fear of falling is an inbuilt response in many animals. In a famous series of experiments carried out in the early 1960s, American psychologists put animals of many different species on a laboratory 'cliff', from which a sheet of glass was extended so that the animals could, if they chose, walk safely past the precipice without falling. All of the animals tested, except for some flying and swimming species, shrank back from the sight of the precipice. Goats behaved that way on their first day of life; human infants did so as soon as they could crawl.

In small chambers inside your ears you have chalky stones called otoliths, which fall downwards towards the Earth until they are brought up short by hairs lining the chambers. Nerves at the base of the hairs respond to the pressure of the stones by sending signals into the brain saying 'this way is down'. Tip your head sideways and a different set of hairs will be stimulated by the stones. Your sense of downness is always with you—unless you are in a spaceship.

Real-life astronauts in space laboratories—the American *Skylab* and the Russian *Salyut*—have endured for months on end the peculiar life-style known variously as 'weightlessness', 'zero-g' or 'free fall'. While their craft orbited a thousand times around the Earth, they could cavort like no acrobats before them, and sleep in any position or attitude. Spilt liquids hung in the air in blobs, loose objects drifted through the cabins and the astronauts had to use vacuum razors and vacuum toilets, lest the refuse should float intolerably around them. The men became physically taller, and their bones weaker. That weird life is the epitome of einsteinian gravity. A spacecraft with the engines switched off is in a state of free fall and *no force* acts on it or on its contents. This is despite the fact that the spacecraft may be seen whirling around the Earth just as if it were on the end of a taut string.

When Jules Verne, in the nineteenth century, imagined people flying from the Earth to the Moon in an enlarged artillery shell, he had the wit and humour to envisage weightlessness. But he made its occurrence and the reason for it entirely wrong, in a way that helps to expose a clumsiness in pre-einsteinian thinking about gravity. For Verne weightlessness and the associated hi-jinks set in for only a short part of the trip—when the gravity of the Moon exactly balanced the gravity of the Earth. But we know—not just by reasoning but by experiment—that the idea is false. In the real-life *Apollos* going to the Moon it was weightlessness all the way—just as in *Skylab, Salyut* and all other spacecraft, once the rockets have stopped firing.

Isaac Newton would not have made that error. He would have recognised that the lunar spaceship and its contents were falling freely—yielding completely to the force of gravity, because they were slowing down as they climbed away from the Earth. The travellers in Verne's shell would have been no more aware of the pull of the Earth in their wake than we are of the pull of the Sun on the Earth. Yet Newton's idea of forces reaching out like invisible grapnels from Earth and Moon was what prompted Verne's error.

In newtonian physics you first invoked the force of gravity and then magicked it away when objects yielded to it. That required quite a rigmarole, in principle if not in practice. You found the mass of the Earth and an object affected by its gravity—let us say a book drifting loose in the spaceship. (If I allow myself a dash of mathematics here it is only to show how complicated newtonian gravity was.) You measured the distance from the centre of the Earth to the book. Then you said that a force acted on the book proportional to the masses multiplied together and divided by the square of the distance between them. Next you had to apply that force to the mass of the book and compute its trajectory within the Earth's gravitational field. In fact you had to repeat the operation, individually, for the spaceship, the astronauts and every loose manual and razor-trimming aboard it. Lo and behold, they all accelerated together at the same rate, creating an impression of weightlessness. It's a wonder each of those pieces did not need a computer to figure out what to do next. By the new common sense of the space age, does it not seem much simpler and nearer the truth to say, with Einstein, that no force acts on the weightless spacecraft—or on his legendary man falling off the roof in Berlin?

The old common sense about motion stemmed from everyday experience. People walk where they please. They steer bicycles or jet planes to any chosen destination. They see water meandering about a valley and feel the air moving in winds from any quarter. Birds and fishes criss-cross the oceans with their great migrations, spiders and sailors climb nimbly up and down ropes, and spaceships clamber right up to the Moon. Everything, it seems, travels independently in its chosen direction and at its own chosen speed. But that is a false impression: the view, if you please, of mice cavorting freely in a cage in the hold of an air freighter.

Take any or all of the travellers mentioned to a precipice on the airless Moon and bulldoze them over the edge. They will then immediately head in the same direction, downwards, travelling in company just like objects in a weightless spaceship. Gathering speed at the same rate they will hit the ground at exactly the same moment. It makes no difference whether they are conscious or unconscious, struggling or passive, heavy or light. They all ride together to their doom on an invisible public escalator created by the Moon's gravity. Galileo predicted it, Newton puzzled over it, but only in Einstein's account of gravity did it become plain why everything falls at the same rate in a vacuum.

Meanwhile the Moon itself moves on its regular way around the Earth, like a tram on an invisible track. The Earth and its attendant Moon orbit the Sun together, and the Sun and all its minions travel at 175 miles a second around the centre of the Galaxy. The Galaxy itself is hurtling under the influence of other galaxies. Our optional random movements on the Earth or Moon are utterly trivial compared with our shared trajectories in the gravity of cosmic space. To depart from them is far from easy: around 1970 going to the Moon cost about a billion dollars per traveller. Gravity rules the universe, but according to Einstein it exerts its authority not with force but with persuasion—the ineluctable persuasiveness achieved by laying down convenient tracks through space and time.

A preference for the old newtonian or the new einsteinian theory of gravity is not a matter of taste or intuition but of experiment. Although they agree almost exactly about the action of gravity in mild conditions, they take completely different views of the phenomenon and in extreme conditions the theories flatly disagree. Moreover Einstein's theory predicts effects undreamed of in Newton's philosophy. For practical purposes many people still employ newtonian formulae, much as an atheist will borrow biblical phrases. But Newton's theory is definitely disproved and by all the available evidence General Relativity is correct.

Even the most astute reader is unlikely to grasp the einsteinian interpretation instantly. But a brief summary of some of the ideas of Einstein's theory of gravity may be helpful at this stage, if only to itemise the unfamiliar points that call for further explanation, elaboration and verification.

Gravity slows down time. Clocks on the surface of the Earth or the Sun run less energetically (more slowly) than other clocks run, farther out in space. An object such as an apple at rest on the ground is also seen to have less energy in the region of slow clocks than it has when at rest farther out in space—up in the tree for example.

An apple falling freely has no force acting upon it. Consequently it can neither gain nor lose energy. But if it enters a region of slower time you will see it losing energy unless it acquires energy of motion. Accordingly you see it speed up as it falls, and its total energy remains unchanged. Then it hits the ground—splat! It sheds its energy of motion and settles down with the lesser rest-energy befitting its lower station.

The full story is not quite as simple as that. The speed of light provides a fundamental connection between distance and time. Clocks cannot go awry in this manner without affecting space. Even light travels along a curved path. The distortions of space and time reinforce one another and create the invisible tracks on which unpowered objects run—for example the orbit of the Moon around the Earth. In short a massive body distorts time and space around it, and those distortions guide the movements of other objects in its vicinity. Gravity is a peculiarity of space itself, not of individual items in it.

The reiteration of 'clocks' and 'time' in this summary of Einstein's theory of gravity is bound to provoke in the reader's mind fascinating and legitimate questions about the nature of time. In my experience these questions are often a stumbling block in trying to understand relativity; it is like being distracted from an explanation of how an engine works by worrying about the pollution that it causes. I shall leave the philosophical issues aside until I have sketched the argument of General Relativity and told how its predictions have been, or are being, tested in practice.

Helpful to this strategy of postponing the general questions about time is the fact that the most appropriate timekeepers for the purposes of the argument are atomic clocks. The reader is likely to be open-minded about how atoms behave and how their timekeeping operations might be influenced by gravity. Just how deeply the variations of atomic clocks cut into our preconceptions about time in general, and the human experience of time in particular, can then be treated as a separate issue.

7 : Shells of Time

Atomic clocks run slower at ground-level than up in the air.
Gravity can arrest light and create a black hole.
Time stands still at the edge of a black hole.
Slowing down time reduces the energy of light and matter.
A falling apple loses rest-energy and gains energy of motion.

A US Navy aircraft flew slowly in a tedious 20-mile loop around and around Chesapeake Bay for fifteen hours on end. It was carrying out one of the classic experiments on the nature of gravity and its effects on time, as predicted by Albert Einstein; 'classic' in the sense that the measurements finally laid to rest any reasonable scepticism about gravity influencing the rate of clocks. Indeed the distinguished theorist John Wheeler compared this experiment with Galileo's, which in 1590 disposed of the lingering belief that objects fell at different rates. It was in 1908, some years before he perfected General Relativity, that Einstein advanced the idea that gravity should affect clocks. Yet even in the 1970s some critics were still trying to deny it. The Chesapeake Bay experiment, involving five flights between September 1975 and January 1976, ought certainly to silence them for all time.

The operation showed, very simply, that time depends on where you are and clocks run more slowly at ground level than they do in an aircraft flying in the weaker gravity high above the Earth. Carroll Alley of the University of Maryland masterminded the experiment, which used two 'ensembles' of very precise atomic clocks. Each consisted of three caesium and three rubidium clocks carefully protected against vibration and any changes in pressure, temperature and magnetism. During each flight one ensemble of clocks went in the aircraft, while the other stayed on the ground.

Laser flashes every three minutes or so compared the times recorded in the air and on the ground, in the manner prescribed by Einstein himself for the comparison of clocks. While the aircraft circled at an average altitude of 30,000 feet (roughly the height of Mt Everest) the clocks aboard it gained about three billionths of a second every hour. Radar tracked the aircraft continuously so that allowance could be made for its motion: for reasons to be explored in a later chapter, the aircraft's speed tended to *slow* the clocks a little. The increase in clock-rate due to height confirmed the einsteinian prediction to within about one per cent. It was no illusion: the effect did not disappear when the aircraft landed. After each flight the clocks that had flown were compared with the ground-based clocks and were plainly about 50 billionths of a second *older*.

In June 1976 a conceptually similar experiment began from Wallops Island, Virginia. A Scout rocket climbed away from the Earth, carrying a single hydrogen-maser atomic clock provided by Robert Vessot of the Smithsonian Astrophysical Observatory. The clock rose to an altitude of some 6000 miles before falling back into the Atlantic Ocean. Radio signals from the clock were automatically corrected for the grossest effects of its speed in outer space. Once again the rate of the clock increased as gravity weakened its grip at high altitudes. At the top of the path, 6000 miles above the Earth, the clock was running 'fast' by almost one billionth of a second every second—a discrepancy about 700 times greater than in the Chesapeake Bay experiment, just as Einstein would have expected.

The mad hydro-electric engineer in an earlier chapter made the Earth into a black hole by squeezing it into a space less than an inch across. In Einstein's theory the main features of gravity around the Earth are exactly the same as they would be if our planet were hollow, with just a faked-up *papier-mâché* surface of mountains and seas, but having a black hole at the centre with the same mass as the Earth. Beneath the surface the comparison does not give the right answers, but that is by the way. On the surface, and out into space, it is correct. This is the model of the Earth that I shall use in what follows.

The German theorist Karl Schwarzschild offered this very useful interpretation almost immediately after Einstein published his theory. In doing so, Schwarzschild became the begetter of the modern idea of black holes, half a century before astronomers took any serious interest in them. Whether black holes exist or not is no more relevant to this argument about the Earth's gravity than asking Euclid if perfect triangles exist. The black hole is for reasoning with. It provides the quickest and most graphic route to the essential points about Einstein's theory. To avoid any misapprehension, let me emphasise that Schwarzschild's black hole at the Earth's centre is a mathematical fiction.

All that is required to create a black hole, in the imagination at least, is that light should feel the effects of gravity. Even in the eighteenth century a great French newtonian theorist, Pierre-Simon de Laplace, imagined that a very massive star might choke off its own light, by stifling it with intense gravity. When Einstein figured out that light had mass—according to the formula $E = mc^2$—it was inevitable that light should be influenced by gravity as surely as any overweight human being or a stone flung into the air. Sufficiently intense gravity should then create a black hole—defined as a region in space from which even light cannot escape.

Faraway objects tend to fall towards a black hole just as they do towards a planet or a star. Like the Earth or the Sun, the black hole has a centre, a certain size and a round shape, so that you could fly around it and inspect it from any direction in space. What distinguishes a black hole from a planet or an ordinary star is that anything falling into it cannot come out of it again. If light cannot escape, nothing else can and it is a perfect trap: a turnstile to oblivion. The trap is not at the centre of the black hole, but at some distance from it.

In the case of a black hole as massive as the Earth that distance is a third of an inch; for the Sun the corresponding radius would be almost two miles. That is the radius of the deadliest sphere of influence surrounding the centre. Physicists call the surface of this special region the 'event horizon' because, if you are looking at it from the outside, no events can be seen beyond it. There are no material markers like the rocks and oceans of the surface of the Earth: the black hole's surface exists in empty space. But if you cross it your body will never be recovered.

Picture now a brave little particle of light on the very edge of the black hole, heading outwards and trying to escape from it. It has stopped, being stuck on the very borderline between success and failure like a fly on flypaper. If gravity affected light in essentially old-fashioned ways, the fly would be struggling: you would picture the particle of light puffing away, losing energy with every passing moment. You might even be tempted to imagine the light dropping back into the black hole, like a spent rocket. But in fact the particle of light is entirely relaxed, neither gaining nor losing energy, not advancing or falling back. A fundamental difference between Einstein's and Newton's view of gravity is that, because gravity affects light, it also affects time. There is no 'passing moment' because, like the outward-bound light, time itself stands still on the edge of a black hole.

Light in the broadest sense and the frequency of its vibrations serve for timekeeping in the very best clocks available to us—atomic clocks. Moreover, light travels rapidly through space, carrying its vibrations with it. That means you can easily check the running of an atomic clock that is far away. You can call up the attendant and ask him to flash a light at you, once a second, as measured by the pulse-beat of his atomic clock.

Now imagine an astronaut stationed at a safe distance from a black hole, expecting a signal from an atomic clock at the black hole's surface. He would wait in vain, because by definition no signal can ever reach him. He is entitled to conclude that the clock has stopped. But he might also suspect that the clock had been swallowed by the black hole, so it is more helpful to consider an atomic clock just outside the black hole. The effects of gravity on time do not suddenly 'switch on' at the surface of a black hole but develop gradually as you approach it.

An atomic clock close to a black hole runs slowly, as monitored by the distant astronaut. Suppose, for example, that it is designed to flash a white light every second. Two things will happen to it, as judged by the astronaut who is equipped, we can suppose, with an identical brand of clock. The flashes from the clock near the black hole will come less frequently—say, one every two seconds instead of once a second. Secondly, the light will appear red instead of white. The colour of light depends on its frequency and, just as the frequency of the one-second flashes is halved, so is the frequency of the light itself. White light is a mixture of frequencies, with blue light having twice the frequency of red light. The red light becomes invisible 'infra-red' rays and the blue light becomes red light.

When Einstein declared that light consists of particles, he also stipulated that the energy of each particle be proportional to the frequency. Therefore, from the astronaut's point of view, the particles of light coming from near the black hole have lost energy. Now if a newtonian physicist had known, as Einstein did, (a) that light is heavy and (b) that its frequency depends on the energy of its particles, he could have predicted the change in colour. Climbing away from the black hole, like a rocket fighting the newtonian force of gravity, the light would lose energy and gradually become reddened in the process.

The old and the new theories of gravity agree about that, but the difference in interpretation is crucial. Recall the legendary man falling off the roof in Berlin. Einstein said that he felt no force of gravity. Similarly, light climbing away from the vicinity of a black hole feels no force of gravity. It cannot lose energy en route. If, therefore, the light appears reddened it is because it *starts off* reddened, because the atoms and atomic time run slow near the black hole. The newtonian analysis would not predict that the light flashes—the one-second light flashes from the atomic clock—also arrive less frequently at the astronaut. For Einstein the slowdown of light frequency and the slowdown of clock frequency go hand in hand. And that means, as we shall see, that natural atomic 'clocks', provided by the normal activities of atoms and the associated emissions of light, will in principle serve just as well as manmade clocks in testing the predictions of Einstein's theories.

Next, imagine stationing the flashing atomic clock at increasing distances from the black hole. It will run faster and faster at each stage: the flashes will become more frequent and the light will appear less reddened. The black hole is, in effect, surrounded by successive zones, or 'timeshells' as I shall call them. They are like the layers of an onion. In different timeshells, atomic time proceeds at different rates.

What is the rule about the rate of clocks? The plainest way of describing it is to call the rate of the distant astronaut's clock 100 per cent, and the 'slowdown' is then a percentage reduction of the clock rate; a one per cent slowdown, for instance, means that the clock is running at 99 per cent of the 'standard' rate. That degree of slowdown occurs at a distance from the black hole's centre 25 times greater than the diameter of the black hole. In the case of the black hole of the mass of the Earth, that corresponds to about 18 inches. You have to be quite close to the black hole for the slowdown to be as great as one per cent.

Nearer to the black hole, the rule about clocks becomes a little complicated, but at greater distances it is about the simplest imaginable: double the distance from the black hole, and the slowdown of the clock halves. For example, the slowdown at 36 inches from the earth-like black hole is half of one per cent. At the Earth's surface, 360 million black-hole diameters from the centre, the slowdown compared with the distant astronaut's clock is about 14 parts in a billion, or about half a second a year. Go 4000 miles out in space and the slowdown halves again. The differences in clock rates are very slight, yet they keep us pinned to the Earth. Our *papier-mâché* Earth's surface, put in place around the black hole, fits roughly into a single timeshell, although atomic time proceeds a little faster at the top of the Himalayas than at the bottom of the ocean. The experiment with airborne atomic clocks, described at the beginning of the chapter, was exploring and verifying the fine details of the Earth's timeshells.

Our astronaut is still a long way off in space, but now he looks at light coming from the surface of the Earth. Although the slowdown in frequency is much less marked than in the case of the clock near the black hole, a slight reddening effect still occurs. Called the gravitational redshift, or the einstein redshift, it is similar to the doppler redshift which would occur if the astronaut were travelling away from the Earth, but the reason for it is different. The prediction of the gravitational redshift is a basic feature of Einstein's theory of gravity and physicists and astronomers have tested it very carefully. They do not have to leave the Earth to verify the effect, because time runs detectably slower at the bottom of a high tower than it does at the top.

In experiments at Harvard in 1959 and 1965, physicists measured the effects of the Earth's gravity on particles of light—on gamma-rays emitted by the nuclei of atoms, to be precise. Robert Pound and his colleagues were able to detect the apparent loss of energy, or 'reddening', of the gamma-rays as they climbed seventy-four feet up a shaft running from the basement to the penthouse of the Jefferson laboratory tower. The effect was very small, but a technique of high precision had recently become available. A young German physicist, Rudolf Mössbauer, had discovered that the nuclei of atoms embedded in a crystal can emit and absorb radiation of very precisely defined frequency—they are, in effect, 'nuclear clocks'.

In order to synchronise two such 'clocks', at the bottom and top of the tower, so that one would absorb the gamma-rays emitted by the other, the Harvard experimenters found that they had to move one or the other of the clocks at a small speed. In other words, they used the doppler effect to compensate for the difference in frequency in the gamma-rays, and so make the 'clocks' match. By this means they were able to measure a discrepancy of a few parts in a million billion, and to confirm the gravitational redshift to within one per cent of the einsteinian predictions.

Gravity is stronger on the Earth's surface than it is out in space, but on the surface of the massive Sun it is stronger still. The view of the Sun from the Earth is thus similar to that of the astronaut looking at the Earth. Atomic clocks on the Sun will seem to run slower than clocks on the Earth, by about a minute a year. Light from identified events in identified atoms in the Sun turns up in telescopes on the Earth slightly reddened by the gravitational redshift. In 1962, French astronomers announced that they had detected the effect.

In 'white dwarf' stars, which have collapsed in their old age, gravity is very strong; on the surface of a white dwarf you would find yourself weighing thousands of tons. The reddening of the light due to gravity should therefore be correspondingly marked, although still far too small to be noticeable by eye. Using instruments, astronomers have detected a clear gravitational redshift in white dwarfs. There are difficulties in disentangling the reddening of the light due to the gravitational redshift from the doppler effects of motion and rotation of the stars. But in 1977 the distinguished American astronomer Jesse Greenstein published, in collaboration with Alec Boksenberg and his colleagues from University College London, a careful analysis of the spectra of light from more than a dozen white dwarf stars, made with the 200-inch Palomar telescope. The average effect was equivalent to an apparent slowing down of atomic time in the vicinity of the white dwarf by more than an hour a year.

The timeshells around a massive body are therefore real, as judged either by the reddening of light or by atomic clocks. Back on the Earth, time passes more slowly on the ground than it does at the top of a tower —or the top of an apple tree. We are now able to see what happens when an apple falls to the ground. Newton said it was because a force of gravity pulled it, causing it to accelerate downwards. The einsteinian account is quite different.

Atoms, atomic clocks and light all vibrate slowly in regions of strong gravity. Because of the link between energy and frequency, they possess less energy than they would do in space, a long way from the source of the gravity. An apple consists of atoms and possesses less energy when lying on the ground than it has when on the tree. Its rest-energy, in the sense of Einstein's equation $E = mc^2$, is reduced.

Picture the apple hanging from the bough. The stalk breaks and the apple falls towards the imaginary black hole at the centre of the Earth. In doing so it enters shells of slower time. As far as the apple is concerned, no force acts upon it. And yet it plainly gathers speed as it plunges. Why?

As no force acts on the apple, it cannot gain or lose energy. But it loses rest-energy just by entering a region of slower time. It must therefore exhibit some other form of energy to maintain the total. The apple cannot maintain its energy by changing colour or humming a tune. The only way it can compensate for its loss of rest-energy in the slower timeshells is by piling on energy of motion—by moving faster and faster. If you reckon how it must gather speed to keep its total energy constant, it has to accelerate steadily—at 32 feet per second per second, just as Newton had it.

In saying that the rest-energy plus the energy of motion remains unchanged for the falling apple I am simplifying a little. (In the relativist's language, the quantity that remains unchanged is actually 'the scalar product of the tangent vector with the Killing vector'!) The complexities of a fuller account arise from the generality of General Relativity and its meticulous attention to who is measuring what energy by which clock. But the version I have given here captures, I believe, the essence of what is going on, in the gradual release of the rest-energy of ordinary matter as an object falls closer to a black hole—or to the Earth.

8 : Directed Futures

Light is heavy and gravity bends its path.
The bending of light means that gravity warps space.
Unpowered objects travel as straight as they can through warped space.
Gravity acting on light gives time a direction in space.
A spinning massive object drags space around with it.

How does the falling apple know which way to go, without any 'force of gravity' to guide it? It has no scouts out, measuring time-rates in different directions. It does not even have ears with sensors, like ours, telling it which way is 'down'. Still less does the apple have a PhD in relativity theory. But it moves off unhesitatingly on the invisible escalator that leads straight down towards the centre of the Earth. The explanation lies in the way a massive body like the Earth distorts space as well as time. The distortions of space are usually described by saying that space is 'warped', or 'curved'.

The simplest meaning of that statement is that light does not travel along straight lines. Because light has energy and is subject to the effects of gravity it will, in Albert Einstein's theory, tend to fall towards the Earth like any other object. Light goes so fast that this tendency is hard to spot. For example, a laser beam fired at the Earth's horizon will drop only a third of an inch in four thousand miles before rushing on into space. Nevertheless this behaviour severely wounds the old notions about geometry, 'the measure of the world'.

Close to a black hole the curving of the path of light is much more marked and you can see around corners. Objects judged in the ordinary way to be behind the black hole, and eclipsed by it, will be visible out to one side. Our conventional ideas about space rely upon light travelling in straight lines. For example we can tell whether a wall is straight by squinting along it, but near a black hole a wall built on that principle will be bent. The traditional rules of geometry no longer apply. It would be easy if one could say that the path followed by light defines the lines of curved space. Unfortunately that is not the case, in Einstein's theory. Half of the bending of light is due to the effects of gravity on time, so the curvature of space is only half of what you might judge it to be by watching a ray of light.

Another symptom of the curvature of space is that the surface area of a massive body is less than you would deduce by measuring its radius and applying the euclidean formula ($4\pi r^2$). In other words the radius is too great for the area, and the 'excess radius' is proportional to the mass of the object. In the case of the Earth the excess radius is about six hundredths of an inch, which means that its surface area is about sixty acres less than you would expect. For the Sun, the excess radius is about 500 yards and the 'missing' area is about 3·4 million square miles—roughly the area of the USA.

A model of space made from a rubber sheet provides a customary illustration of curvature of space. The sheet is stretched flat like a trampoline and then weights are attached to it, to represent stars, planets or black holes. The weights deform the rubber sheet downwards around them. Then you say that in this model the flat rubber sheet is a collapsed, two-dimensional representation of three-dimensional space, and the indentations symbolise the curvature of space.

Such a model helps the imagination. If you represent a spaceship or a moon by rolling a small ball-bearing on the rubber sheet, it will follow a curved path, bending in towards the massive objects, like the real objects moving under gravity. It also gives, correctly, an impression of space being stretched along radial lines running straight towards a massive object—creating the 'excess radius'. In the case of a black hole the deformation becomes enormous. But there are two drawbacks. One is that the model represents only half of what is going on: it does not show the slowing of time.

The greater drawback is that space is not a flat rubber sheet and it is impossible to make the mental transition from the sheet to real, three-dimensional space which is also curved. Don't try! Even mathematicians find it difficult to visualise curved space. Nevertheless it is comparatively easy for relativists to reckon the distortions and to measure them, without attempting to make pictures in their heads. The rubber sheet is as good an aid to thought as any.

The 'light-bubble' is another aid, though, and it gives a clearer sense of how time and space become distorted and churned together near a black hole and other massive objects. Imagine a spaceship with beacons arranged to give off light in every direction. The spaceship is then like a miniature star. Light leaving the spaceship at one instant will, in the next moment, reach a certain distance out from the spaceship, so forming a bubble of light surrounding the spacecraft. For example if the 'moment' is a millionth of a second, the diameter of the light-bubble will be roughly 660 yards; of course, as the light rushes on its way in all directions, the light-bubble is actually expanding at the speed of light. If the spaceship is in empty space, far from any massive bodies, its light-bubble will surround it equally: the spaceship lies dead-centre in the midst of the light-bubble. But even in this simple situation the light-bubble has a profound meaning. It defines all possible futures for the spaceship and any astronaut it carries. Because they will never be able to travel faster than the light, they can never move outside their own light-bubble. For example, if the astronaut wants to visit a star ten light-years away, he cannot reach it in less than ten years, which is how long it will take for the light-bubble being puffed out *now* to grow to include the star. Any proposal for being at that star next Friday falls outside the astronaut's light-bubble, and is simply inaccessible to him.

Now put our spaceship near a massive object. Gravity displaces the light-bubble off-centre and as a result the astronaut's future is somewhat biassed in a certain direction in space—towards the massive object. An interchange occurs between time and space. Its meaning assails our prejudices about time more fiercely than the slowing down of clocks. A black hole again clarifies the point. Imagine the spaceship just crossing the dire perimeter at the surface of a large black hole. The astronaut is trapped for ever. His future now lies inside the black hole, because the light constituting the light-bubble cannot, by definition, escape from the black hole. By no exertion can the astronaut or his spaceship manage any better than the light. But to speak of the light-bubble in this connection not only reconfirms in different words the deadliness of the black hole: it reveals the interchange between time and space.

The light-bubble of the doomed spaceship expands, but only within the black hole. While outward-heading light is stopped, ingoing light can still travel rapidly towards the centre of the black hole. The result is that the spaceship at the edge of the black hole is completely off-centre with respect to the light-bubble, which is now biassed wholly inwards. Bearing in mind that the light-bubble defines all possible futures of the spaceship and astronaut, you are led by this entirely straightforward reasoning to a remarkable conclusion: that *time* for the

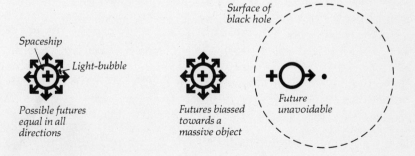

Spaceship

Light-bubble

Possible futures equal in all directions

Futures biassed towards a massive object

Surface of black hole

Future unavoidable

astronaut has been converted into a certain direction in *space*. All the future that he has lies towards the heart of the black hole.

Even the Earth with only a very small black hole at its centre in imagination, and none in reality, has similar though much milder effects on time and space. Common sense tells us that our own future time lies towards the Earth rather than in space, unless we make prodigious efforts to escape from the Earth's gravity. That bias is officially incorporated in Einstein's theory. The future course of time is not equal in all directions at the surface of the Earth. It takes on a direction in space going towards the Earth's centre. And that is how the mindless apple knows which way to fall. The direction imposed upon the apple's future —revealed by the slight displacement earthwards of the apple's light-bubble—means that the laziest thing that it can do is to travel that way. Some effort would be needed to make it follow any path that is not straight down; in its natural fall, no force is involved.

The imaginary black hole at the middle of the Earth ought really to be spinning on its axis, just as the Earth does. When you bring rotation into General Relativity, new effects appear, unknown in Newton's gravity. Once again the black hole makes the point with more vividness than does any subtle reasoning about the real Earth. A rotating black hole is slightly more elaborate than one which is not rotating. Outside the surface of the black hole there is a buffer zone called the 'ergosphere'. It is a sphere flattened at the poles of the black hole's axis, where in fact it touches the surface of the black hole.

The ergosphere is a nightmarish carousel, a region of space where any object is forced to whirl around the black hole at high speed. In principle you can step off the carousel: light can still escape from it, and indeed this is the zone exploited in Roger Penrose's fictional machine for extracting the rest-energy from garbage (Chapter 4). But, while you are within the ergosphere, revolutions are obligatory. This is despite the fact that the ergosphere consists of nothing but empty space.

Gravity performs this trick by the churning together of time and space, but now in a way that makes the future time of everything lie, not towards the centre of the black hole but 'sideways', around the carousel. The light-bubble of any object is displaced completely in the direction of rotation. In other words, light is forced to travel in that direction and there is literally 'no future' in trying to swim against the stream, to go around the other way or even to stand still; to do so, you would have to go faster than light. Because of this intrusion of time into the directions of space, the rotating black hole drags space and everything in it around itself at high speed.

The carousel is an extreme situation, but the effect does not suddenly switch on: even at a distance from the rotating black hole future time has some bias in that 'sideways' direction. Under this kind of influence, a little removed from the black hole, you would feel giddy and see the distant stars apparently whirling around you. Even at the Earth's surface, with the imaginary black hole at the Earth's centre rotating once a day towards the east, the eddying of space should persist. The effect is very small, partly because, by black-hole standards, the rate of rotation is very slow, and partly because we are 4000 miles from the Earth's centre.

If the Earth is in this sense dragging space around with it as it rotates, the effect should be detectable with a high-quality gyroscope spinning in a spacecraft that is orbiting the Earth. The axis of the gyroscope ought to swivel at a very slow rate, corresponding to one full turn in twenty-five million years. There is another effect, causing the gyroscope to swivel about a hundred times more rapidly; it is due to the spacecraft's own motion through the curved space around the Earth. But experimenters are hopeful that both effects will be detectable, with the help of modern precision engineering. In 1979 a gyroscope experiment of that kind was being prepared by a group at Stanford University in California, to be put into orbit by the Space Shuttle in the early 1980s.

The gyros for this extraordinarily delicate experiment will take the form of spheres of quartz, about the size of table-tennis balls. They have to be almost perfectly round, to better than a millionth of an inch. Each is coated with a thin film of a metal that becomes 'superconducting' within a few degrees of absolute cold, allowing the gyro to be suspended in a magnetic field. Novel techniques for setting the gyro spinning and keeping the track of its axis of spin have been developed at Stanford. The envisaged satellite will carry four such gyros; in addition another ball will serve as the 'conscience' of the satellite itself. Satellites are normally buffetted by the pressure of sunlight, for example, so that they do not execute a perfectly gravitational orbit; to avoid this the 'conscience' detects any slight departures from free fall and corrects them by releasing jets of gas. To provide an axis of reference for the satellite, against which to compare the motions of the gyros, the satellite will automatically keep pointing at one bright star.

It adds up to heroic engineering, pursued at Stanford and associated laboratories since 1963. You may wonder why a single experiment is worth so much effort, just to measure the dragging of space—an effect that is far too small to be of any practical consequence. A general answer is that the effort is symptomatic of the importance that present-day physicists attach to Einstein's interpretation of the workings of the universe. More particularly, the dragging of space by the rotating Earth, and the consequent swivelling of a gyro near the Earth, represents not a modification of Newton's ideas but a brand-new effect in einsteinian gravity. Among other benefits this promises a clear-cut result. The experimenters will not have to fish—as is often the case—for subtle einsteinian effects among gross and complicated changes that have nothing to do with General Relativity. Either the gyros will swivel at the right rate, or they won't.

But what is going on? What kind of thing is time, if it can be slowed and directed by gravity? What is space, that a massive body can mis-shape it, as if it were a piece of putty, or spin it incessantly around a rotating black hole without tearing it? No simile is really apt because time is time and space is space and they are not like material objects. Yet time and space participate actively in the workings of the universe. And the light-bubble shows how action, and especially the action of light, maps out space and time.

Einstein took some trouble to explain why his kind of space is so different from Euclid's and Newton's. He said that you could imagine a universe in which space was not warped, but it would be completely devoid of matter. Empty space has no practical meaning: space cannot exist separately from 'what fills space' and the geometry of space is determined by the matter it contains. For some readers, the explanations of the master should be sufficient; for others, I shall attempt a modern elaboration.

The way in which time, space and the speed of light are linked together is not mysterious in principle, even though its consequences are weird. Any speed is *defined* as a certain distance travelled through space in a certain time. But in Einstein's universe the speed of light is more fundamental than space or time. Space is what light moves in; time is how long it takes to move. It is by means of light that we see objects scattered through space and changing in time. Atoms, too, rely on light to inform them about what is going on. As a result all physical processes are governed by the speed of light.

Even more fundamental than the speed of light is energy. The accumulation of matter and its rest-energy in a massive body affects light by way of gravity. Effects on the behaviour of light have consequences for the reckoning of space and time. A black hole, for example, warps space while it slows down the passage of time and, at short ranges, arrests it, as judged by a distant onlooker. He will also see light being delayed near a black hole. One way of explaining that is to say that distances are stretched, as in the rubber sheet described earlier, and the light has farther to travel at its usual speed. Another, and simpler, way is to allow that light slows down from the onlooker's point of view.

So think of space as a peculiar kind of glass, which is transparent to solid objects as well as to light. Glass slows light down and the denser the glass the slower the light. A massive body squeezes and deforms space in its vicinity. Space is 'denser' in the vicinity of a massive body and so light seems to travel more slowly there, when seen from afar. That makes the space act like a magnifying lens, and light passing the massive body bends somewhat inwards towards it. But locally—that is to say, to a person or atom on the massive object or to a particle of light passing by—the speed of light and the rate of time seem to be the same as usual. Accordingly distances must shrink to match the slowing of the clocks.

The 'glassy' space acts also like a valve. Light travels more readily and faster towards the centre of gravity than away from it: the light-bubbles have illustrated this point. (Incidentally, there are real crystals through which light travels at different speeds in different directions.) Accordingly time also passes more rapidly in one direction than in another—time, as we have noted, acquires a direction space. The lazy direction for an object to travel in is the direction in which light and time pass most rapidly—even though it typically leads the object into timeshells where light and time slow down. Any resistance or other exertion on the part of the object to follow a different route involves a slowing down of time.

But the glassy picture of space is too static, and not only in the sense that the massive body itself is actually hurtling through space. Even when pictured at rest in its realm of distorted space and time, the massive body has to work continuously through time to maintain the curvature of space; in doing so it loses a little energy, or mass. Energy oozes out of strongly warped space like water from a squeezed sponge.

Stephen Hawking of Cambridge discovered the process as a remarkable extension of Einstein's theory. Close to a black hole, space is warm and the consequences may be dramatic. Space seethes—that is the reason for the effect. If you take a census of what is present in a volume of high-grade 'empty' space far away from any galaxy, and then discount all the expected things like a few atoms and plenty of particles of light passing through in all directions, something else remains. You cannot detect it in any ordinary way, but one of the strongest theories of modern physics insists that it is there—a surreptitious hint of everything that energy is capable of creating. The existence of ghostly particles predicted by the quantum theory has been confirmed by small effects on the 'tuning' of atoms.

Hawking realised that the intense curvature of space just at the edge of a black hole could convert some of these ghostly particles into durable particles of matter and light. It was in 1974 that he announced that black holes were warm and capable of exploding. The process can run away, draining the rest-energy of the black hole and abruptly abolishing it with a great outpouring of gamma-rays and sub-atomic particles. That would occur in our era only in very small black holes, which might have been created in the Big Bang at the origin of the universe.

Will black-hole explosions be seen? Astronomers might spot the burst of gamma-rays; alternatively, charged particles coming from the explosions will presumably swerve in the magnetism of the Galaxy and so broadcast detectable light and radio waves. Among the experts opinions vary widely about whether the small exploding black holes actually exist; much depends on what you think was happening during the Big Bang. Observations so far set a rough upper limit to their possible occurrence; if black-hole explosions occurred just once a year within a distance of 150 light-years, they should have been seen already.

9 : The Shifting Stars

Light seems to slow down under gravity.
The apparent slowing of light is linked with the slowing of time.
The bending of light means that gravity warps space.
The combined effects on time and space double the bending.
The bending and slowing of light by the Sun are measurable.

Nothing about black holes, exploding or otherwise, was known to Albert Einstein during the ten years 1905 to 1915 when he was trying to make sense of space, time and gravity. Having sampled some of the later ideas and descriptions the reader may be better able to appreciate the nature of Einstein's struggle, while a glimpse of Einstein's own difficulties may sharpen some of the points about space and time rehearsed in the previous chapters. In developing General Relativity he had other important issues to contend with, which will be topics for later chapters, but defining the effect of gravity on light was central to his task.

That there should be some effect was plain almost as soon as Einstein realised that light was heavy—possessed mass—according to $E = mc^2$. One of the most obvious possibilities was that light passing a massive body should be deflected a little by it, much as a bullet is deflected towards the ground. In fact, Einstein had in mind a more profound reason than that for *why* light should fall, but for the moment the question is *how* does it fall? The full effects of a massive body on space and time in the vicinity did not become apparent at once.

Midway along his progress, in 1911, he gave vent to some 'elementary reflections'. By then he had already defined the gravitational redshift and, with it, realised that gravity slows down time. 'Nothing compels us', he affirmed, 'to assume that the clocks in different gravitational potentials must be regarded as going at the same rate.' He also predicted that starlight glancing past the Sun would bend towards it and that this effect could be observed at the time of an eclipse. But his calculation of the deflection was faulty: it was half the correct value that he was to arrive at a few years later.

The master underestimated the effect of gravity on light because he had not appreciated that there are two related and mutually reinforcing effects. One is the influence of gravity on time, which is equivalent to slowing down light and making the space around the Sun into a lens that bends the ray of light. The other is the warping of space which in effect lengthens the path of light approaching and leaving the Sun. That is equivalent to making the lens thicker and more powerful. In 1911, Einstein had the first but not the second effect. It is not a coincidence that the two effects make equal contributions to the bending of light, so that Einstein's early estimate was half of what it should have been. The bending of light by the first effect implies the curvature of space that gives rise to the second effect.

The shortcomings along the way illuminate Einstein's greatness, rather than belittling it. They show him as no mathematical robot coldly summing up the universe for our puny species, but as a man wrestling tenaciously with difficult concepts. The 1911 paper was a hotch-potch of old and new ideas, which is why his calculation of the deflection of starlight was in error. Einstein had still to achieve the grand synthesis which was to feed the idea that gravity slows down clocks into a conspectus for the deformation of space and time by massive objects.

He wanted to understand the universe in the simplest possible way, but his project inexorably drew him into some of the most subtle and mind-tormenting mathematics of his time. The mathematics of General Relativity is complicated because it tries to accommodate all possible situations, in which weirdly-shaped lumps of matter might be passing at high speeds and feeling each other's gravity. I say 'tries' advisedly because, until the advent of very powerful computers, it was impossible to solve the equations of General Relativity in any but the simplest situations. The mathematical tools for describing deformed space and time are not unlike those used for calculating the stresses inside a deformed solid object. The 'tensor calculus' enables the theorist to describe how the stresses of space and time are changing in all possible senses, at any one point. The idea is not mysterious at all, but the mathematics becomes forbidding, and I sidestep it completely.

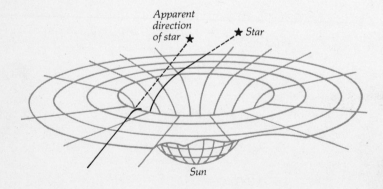

Apparent direction of star ★   ★ Star

Sun

Einstein could not avoid it. He spent a crucial period, 1912–14, with the mathematician Marcel Grossmann in Zurich. Grossmann had coached Einstein in mathematics when they were students together at the Zurich Polytechnic. Now the learned Professor Grossmann taught the not-so-learned Professor Einstein how to play with the geometry of spacetime, using the tensor calculus. Together they used the esoteric mathematics to hunt for the new law of gravity. They almost found it; they wrote it down, but Einstein failed at the time to recognise it for what it was. Still wrestling with the question of what was objective and reliable in the confusing relativistic universe of his creation, he suspected that the laws of nature might vary from place to place.

Not for Einstein the simple arguments about the black hole: he failed to spot that short-cut and anyway the fiction it involved and its lack of generality would have dissatisfied him. Instead he continued on the arduous road, across the Himalayas of apt but difficult mathematics. Further to perturb his mind, he took up a new post in Berlin in 1914, just before war was about to break out in Europe and just as he was falling out with his wife. Yet in 1915, when Einstein was alone with his thoughts in Berlin, the breakthrough came. The laws of nature were the same everywhere. He beheld the beauty of General Relativity, disinterred the correct mathematics, and had his theory of gravity. He adjusted his prediction of the deflection of starlight that would be observable on the occasion of an eclipse of the Sun. As he remarked later:

The years of anxious searching in the dark, with their intense longing, their alternations of confidence and exhaustion, and the final emergence into light— only those who have experienced it can understand that.

The deflection of a ray of light passing the Sun can be reckoned very simply. Imagine that you are standing at the point where the ray passes closest to the Sun. Picture also the imaginary black hole at the centre of the Sun and ask how big it will look to you from that distance. The angle it makes at your eye—that is to say, the difference in direction between opposite edges of the black hole—is the angle by which the ray of light will be deflected. In modern terms, Einstein's correction to his theory doubled the size of a black hole of a given mass.

Although he was in no ordinary sense a vain man, Einstein did not underestimate his own attainments. When he divorced his first wife in 1919, part of the agreed settlement was that he would give her his money from the Nobel Prize. He had not yet been awarded the prize but he was quite sure he would have it before long; in the event, Mileva Einstein had to wait only until 1922 for her cash. At a less domestic level, Einstein was serenely confident that his theory of gravity would be confirmed by observations of the bending of light. In due course a telegram brought the news that it had been done. A girl student asked Einstein what his reaction would have been if his calculations had not been confirmed. He said: 'Then I would have been sorry for the dear Lord.'

British astronomers had seized the first opportunity to test Einstein's theory of gravity, by seeing whether and by how much the Sun deflected the light of distant stars passing close by it. Ordinarily, stars near the Sun are quite invisible because of the glare, but one of the happy coincidences of the universe—ready-made for Einstein, it would seem—is that the Moon fits very neatly across the face of the Sun at times of total eclipses. Then you can see the stars nearly in line with the Sun. If Einstein was right the Sun ought to seem to push the stars outwards from their normal positions in relation to other stars. According to different ways of updating Newton's theory, you could say that the starlight should be deflected either not at all, or by the half-amount erroneously predicted by Einstein before his theory was complete.

A total eclipse of the Sun occurred on 29 May 1919. It was not visible from Europe or America but two expeditions observed it from the Tropics. Einstein's papers had reached England via the Netherlands during the war, and the astronomer Arthur Eddington was at once persuaded of the importance of General Relativity. He was planning the eclipse observations even before the Armistice. One expedition went to northern Brazil, the other to the island of Principe, tucked in the 'corner' of West Africa. At the time of total eclipse the astronomers photographed the stars around the Sun. At first on the spot and then more carefully in England, they measured the positions of the stars and found that the starlight had indeed been deflected. The Principe results showed a deflection slightly less than Einstein predicted; those from Sobral, rather more.

The eclipse results were a triumph. Newton's ideas about gravity had reigned unchallenged for more than two centuries, yet within four years of Einstein developing his theory it seemed confirmed, and Newton was dethroned. The deflection of light by gravity is, as I have stressed, central to Einstein's General Relativity. But later measurements of the deflection of starlight at other eclipses gave a wide scatter of results. They departed from Einstein's predictions by anything up to sixty per cent. The difficulties of the observations were to blame, rather than any defect in the theory. While they did not allow any restoration of Newton they left, nevertheless, a little elbow room for alternative accounts of gravity. So, sixty years after the initial triumph, astronomers and relativists were decidedly cool about this way of checking up on Einstein.

The radio astronomers do better. Quasars are, by definition, pinpoint sources of radio energy, and two of them are particularly well placed for checking the deflection of radio energy by the Sun. They are called 3C279 and 3C273: they lie close together in the sky and on 8 October each year the Earth's motion in orbit brings the Sun into line with 3C279, thus eclipsing the quasar. By Einstein's prediction, 3C279 disappears slightly later, and reappears slightly earlier on the far side of the Sun, than you would expect if it were an inert barrier. Moreover the convenient placing of 3C273 gives the radio astronomers a reference point in the sky for seeing how the apparent position of 3C279 is altered when it is close to the edge of the Sun. By 1970, a number of observations of this annual event, with radio-telescope combinations in the USA and Britain, had confirmed Einstein's prediction to within about ten per cent.

The perfecting of 'very long baseline' techniques, which harnessed together radio telescopes situated thousands of miles apart across continents or oceans, promised a much more precise verification of the deflection. The technical principle here is that the precision with which objects can be pinpointed in the sky depends on the size of the telescope measured in terms of the wavelength of the light or radio waves that it is detecting. You cannot build a complete radio dish as big as the Earth, but you can approximate to it by making simultaneous observations with two or more widely spaced radio telescopes, using atomic clocks to check the simultaneity (defying Einstein, in a sense!). Links of this kind were established between radio telescopes in Canada, the USA, Europe and Australia, and they yielded better knowledge of the small-scale features of quasars and radio galaxies. And applied to the deflection of the radio energy from 3C279 and other quasars the 'very long baseline' technique was expected to verify Einstein's 1·75-second prediction exactly—that is to say, to within about a tenth of one per cent. Meanwhile Einstein was 'scoring' to an accuracy of one per cent, by 1976, in radio observations using a relatively short baseline at the National Radio Astronomy Observatory in West Virginia.

Given the high precision possible with 'transworld' radio astronomy it may not even be necessary to look close to the Sun for the effect of gravity. If you look at the sky just after sunset, the positions of the stars overhead are all pushed very slightly to the east by the bending of their light towards the Sun; at dawn the displacement is to the west. The effect is small—about four thousandths of a second of arc or ten millionths of a degree—but the radio astronomers may be able to pin it down.

In Einstein's full reckoning of the bending of light, the delay to light was implicit. Light passing a massive body such as the Sun should appear to slow down, as judged by a distant observer. No technique was available in 1915 for testing this proposition directly. The difficulty is obvious: in order to measure the speed of light you have to shoot it past the Sun and then get it back again, in order to see how long its journey has taken. In other words, you need an echo. The echoes of radar nowadays make it possible to measure the speed of radio pulses passing the Sun, and radio energy travels at exactly the same speed as light. As judged from the Earth, light indeed seems to slow down near the Sun. Either that, or Venus jumps!

Obtaining radar echoes from the planets was a great technical feat around 1960. The giant radio telescopes coming into service at that time served as the powerful transmitters and ultra-sensitive receivers. When Venus, for instance, is on the far side of the Sun, the radar pulses take almost half an hour to make the return journey, dissipating their energy all the way. Nevertheless the echoes were detected and Irwin Shapiro of the Massachusetts Institute of Technology saw that therein lay a way of checking Einstein's theory of gravity. It is worth recalling that Shapiro the experimentalist encountered opposition from theorists about the interpretation of the einsteinian effect—some denied that any slowing down of the echoes would be observed. This reaction was symptomatic of the misunderstandings of General Relativity, even among the experts, before the present period of experimental verification.

When Shapiro and his colleagues watch the planet Venus by radar as it passes on the far side of the Sun, it seems to lurch several miles farther from the Earth and then to swerve back again, all in the course of a month. The planet does not really do that. The apparent deviation is slight compared with Venus' distance from the Sun (67 million miles), yet you would need a source of energy equivalent to many billions of H-bombs to knock the massive planet off its correct orbit, and the same to knock it back again. No, radar measures the distance of a target by the time taken for the pulses to go out and return. And what the radar astronomers see is a consequence of their pulses slowing down as they pass near to the Sun.

In a classic series of experiments between 1966 and 1970, Shapiro and his colleagues verified that the delays, measured in millionths of a second, were very close to the predictions. For his radar set, Shapiro used MIT's Haystack radio telescope, equipped with a transmitter that generated pulses of 500,000 watts, and a sensitive receiver. He observed both Venus and Mercury and found delays, or apparent 'lurches' of the planets, within about two per cent of what the theory required. The effect is most marked during the month of closest alignment with the Sun, but the cumulative effect over the year is greater, with Venus being apparently misplaced by up to fifty miles altogether.

Similar results, initially of similar precision, were obtained by other scientists at the Jet Propulsion Laboratory in California, in experiments with spacecraft. Signals went from the Earth to *Mariner* spacecraft far off in the Solar System; the spacecraft responded with a signal that was stronger than a radar echo. Again the tell-tale delays occurred, as an effect of gravity doing peculiar things to light and time. More recently, radio exchanges with the *Viking* spacecraft standing on the surface of Mars have achieved greater accuracy. When Mars is lying on the far side of the Sun, the time delays occur exactly as General Relativity requires, to within half a per cent of the predictions. Shapiro continues to take the lead in such experiments, and, he remarks, 'The Solar System, for better or worse, is the principal laboratory for studying General Relativity.'

10 : Tramlines in the Sky

Unpowered objects travel as straight as they can through warped space.
At the right speed an object will orbit around a massive body.
Einstein's description gives answers very like Newton's.
An einsteinian orbit slowly swivels around.
The swivelling of orbits is detectable.

The most majestic sign of the curvature of space is the way the path of the Moon bends into a curve around the Earth, circling us once a month, while the Earth follows a curved track around the Sun, circling the mother star once a year. The planets and their attendant moons seem to run on tramlines in the sky. Our forefathers supposed that gods or angels propelled them; Isaac Newton found that his force of gravity would explain their motions almost perfectly; Albert Einstein abolished the force of gravity and said that the planets and moons were falling freely and travelling as straight as they could go through curved space. As I put it earlier: the massive body distorts time and space around it and those distortions guide the movements of other objects in its vicinity. The curvature is sufficient to cause an object that is travelling at the right speed to go right around the massive body and back to its starting point.

The near-circular paths of the Moon about the Earth and the planets about the Sun do not *look* anything like the track of a falling body. Had the similarity been more obvious Newton's law of universal gravitation might well have been discovered by the Ancient Greeks. It took seventeenth-century genius to realise that the gravity that guided the Moon was the same as the gravity that propelled an apple to the ground.

Newton himself was at pains to help his readers to understand the idea of an orbit under gravity. In the process of explaining it in *Principia* he invented the artificial satellite—although nearly 300 years elapsed before *Sputnik I* fulfilled his idea. Newton invited his readers to imagine a gun on a high mountain, aimed horizontally. A ball fired at low speed will curve downwards in its path and fall to the ground. But at higher speeds, and at long ranges, going beyond the horizon, a new factor comes into play. The Earth's surface curves away from the path of the falling ball, so there is a gain in range. If the gun could fire the ball fast enough, the curving of its path under gravity could be exactly compensated for by the curving away of the ground underneath it.

The ball is still falling, but it does not get any closer to the ground. Instead it circles right around the Earth and keeps on going—in principle for ever because, every time it comes back to its starting point, it is going as fast as before. (This is ignoring air resistance, of course.) The ball has become a satellite. In Newton's theory, the force of gravity acts like a long string, holding the satellite in its orbit like a stone being whirled in a sling. In Einstein's theory there is no force and the satellite runs weightlessly along its natural tramline through space.

So far I have described the distortions of time and space by gravity chiefly in terms of the slowing of clocks, the apparent slowing of light, the displacement of the light-bubble of future time towards a massive object, and the bending of a beam of light that is glancing past a massive object like the Sun. Picturing orbits requires a more general view of the surroundings of the massive object.

Go back to the *papier-mâché* Earth with the imaginary small black hole at its centre, and see how it exerts its mastery over the Moon and artificial satellites. The black hole is surrounded by 'timeshells' where clocks run more slowly as you move towards the black hole. An apple falling straight to the ground enters zones of slower time and speeds up in a manner that maintains its total energy. But in the case of a truly circular orbit, the satellite remains entirely within one timeshell and its energy of motion does not change.

Although it is falling its track is not directed at the centre of the Earth, because of the sideways motion of the satellite. The future time of the satellite has a bias in a certain direction in space by virtue of its motion. Its future is more likely to lie ahead than sideways or behind, because it would need to fire a motor or hit something, to alter its motion.

The track of any object through space is thus a compromise between the 'gravity future', which lies naturally towards the centre of the nearest massive object, and the 'motion future', which lies naturally ahead, along its line of present motion. With no expenditure of energy, and no force acting on it, the object follows a curved track.

The requirement for an orbit is that you should be moving sideways to start with, at the appropriate speed. Go too slowly and you will drop down and collide with the source of gravity; go too fast and you will fly away into space. The speed needed to achieve an orbit around the Earth depends upon the effectiveness of gravity, and accordingly it diminishes as you go out from the Earth. The Moon travels much more slowly than man-made satellites in near-Earth orbit. The slowing down of time, as I have remarked before, halves when you double the distance from the centre of the Earth. The energy of motion needed for a good orbit also halves, and the necessary speed to achieve the appropriate energy is the same for any object. This reasoning about time relates the motion of the satellite to the earlier account of the falling apple.

Looking at the curvature of space offers a neater account of the circular orbit. Recall the flat rubber sheet representing space, and the curving of the sheet produced by fixing a heavy weight to it. You can then think of the Moon or an artificial satellite running around the central mass of the Earth like racing cars on a banked track.

To sharpen this picture, consider the path of light near the Earth; its deflection is the first token of the curvature of space. The bending of starlight even by the Sun is very small; in the case of the Earth it is very slight indeed, yet we can reason with it. At the Earth's surface the path of light travelling horizontally follows a segment of a circle approximately one light-year in radius. You and I could not distinguish that from a straight line; nevertheless it is a symptom of the curvature of space in our vicinity.

Material objects travel more slowly than light and are affected more markedly by the curvature of space near the Earth. Their tracks curve more sharply. In other words they follow a segment of a smaller circle, the radius of which depends on speed. You can then ask at what speed the circle of curvature of the object's track becomes a circle around the real Earth, so that the object goes into a low orbit around it. The square of the speed is what counts and that must be a fraction of the square of the speed of light, in the same proportion as the radius of the Earth is a fraction of a light-year. The necessary speed is just what the Russians selected when they put the first *Sputnik* into orbit: 0·000027 of the speed of light, or five miles a second. Any near-Earth satellite takes an hour and a half to circle the Earth.

What happens farther from the Earth? A light beam passing the Earth at the distance of the Moon is bent far less than light grazing the Earth. At that range the Earth's gravity curves space less. If you reckon the speed that the Moon must possess, for the curvature of its track to keep it at a constant distance from the Earth, it is 0·63 miles per second. That is, naturally, the speed which takes the Moon around the Earth once a month. It also represents just the same reduction in speed, compared with a near-Earth satellite, that would be deduced from the earlier reasoning about time and energy. So to speak of the slowing of time or the curving of space gives just the same answers—as, of course, it must.

Taking stock, we now have the Moon and artificial satellites orbiting in curved space. By travelling as straight as possible and as lazily as they can, they find themselves following circular tramlines appropriate to their speeds and their distances from the Earth. There has been no mention of the mass or composition of the satellite. It makes no difference whether the orbiting object is a natural moon, a piece of machinery like *Skylab,* or a snowball.

Amazingly, Einstein started from a completely different view of gravity and arrived at the same conclusion as Newton did about the relationship between the speed of motion of a satellite and its distance from the Earth. Applied to the planets orbiting about the Sun, it is the law with which Johannes Kepler and Newton conjured when they comprehended the clockwork of the Solar System. Voltaire said of it: 'Nature does not exist; art is everything. . . . Surely some person as clever as the Royal Society in London arranged things so. . . .' Thanks to Einstein we are now privileged to see that, far from being art, it all follows most naturally from the fact that light falls and space curves near a massive object.

So far I have described apples falling straight to the ground, and objects in circular orbits. Real objects tend to be somewhat eccentric in their orbits. The planets go around the Sun in ellipses—squashed circles—rather than perfect circles: that was in fact Kepler's great discovery in the early seventeenth century which paved the way for Newton. During its orbit a planet falls a little closer to the Sun and then turns and climbs away again. As Newton and Einstein give similar answers about straight-down fall and circular orbits it is unsurprising to find that they are in broad agreement about the in-between case of elliptical orbits. And in so far as his theory agrees with Newton's, Einstein can appropriate all the heavenly clockwork that endorsed Newton's theory, in support of his own.

The reckoning of planetary motions is a venerable science. Nowadays it tells us, for example, how gravity causes the ice to advance or retreat on the Earth during the ice ages. The gravity of the Moon and (to a lesser extent) of the Sun makes the Earth's axis swivel around like a tilted spinning top. Other planets of the Solar System, especially Jupiter, Mars and Venus, influence the Earth's tilt and the shape of its orbit, in a more-or-less cyclic fashion, with significant effects on the intensity of sunshine falling on different regions of the Earth during the various seasons. Every so often a fortunate attitude and orbit of the Earth combine to drench the ice sheets in sunshine—as at the end of the most recent ice age, about ten thousand years ago. But now our relatively benign 'interglacial' is coming to an end, as gravity continues to toy with our planet.

On phenomena like these, the predictions of Einstein and Newton concur. If there were no discrepancies at all between the theories it would be a matter of taste as to which account of gravity you preferred. But subtle differences appear, between Einstein's theory and his predecessor's. Turning yet again to an imaginary black hole, for clarification of the differences, recall that the diameter of the black hole in General Relativity is exactly twice what you would calculate by quasi-newtonian rules. The slowing of time and the curvature of space reinforce each other's effect, and the central mass of the black hole can overpower light at twice the range you might otherwise expect. The imaginary black hole at the centre of the Sun has a diameter of 3·7 miles rather than 1·85 miles. At great distances from the Sun, this makes practically no difference to the pattern of gravity. But as you approach nearer to it, the effects of gravity have to increase a little more rapidly than in Newton's theory. A planet close to the Sun has to travel a shade faster in order to maintain a circular orbit. And a planet in an elliptical orbit, diving in towards the Sun, swinging around it and then climbing away, will experience some 'excess' gravity when it is nearest to the Sun. A distant observer sees the planet lingering for a moment in that timeshell, before it moves out again.

In Newton's theory a planet follows exactly the same path for ever except for the calculable perturbations due to the influence of other planets. In Einstein's theory it does not. The planet tends to overshoot its expected 'perihelion'—the point of closest approach where it turns and begins its long climb away from the Sun. Each time it goes around the Sun, the perihelion will inch forward a little. The whole elliptical orbit gradually swivels, or 'precesses', as the centuries go by.

The progression is extremely slow. In the case of the Earth, which approaches most closely to the Sun each January, this 'advance of the perihelion' is only a few miles in an orbit nearly 600 million miles in

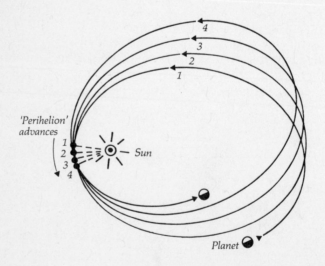

circumference. But Einstein had good reason to emphasise this subtle distinction between Newton's theory and his own. The effect is most pronounced for the innermost planet Mercury, and nineteenth-century astronomers had spotted a discrepancy in Mercury's motions. They tried to explain it by the presence of an unseen planet Vulcan—but when Vulcan failed to show up some dared to suggest that Newton's theory might not be strictly true. Einstein's theory gave a complete explanation of the discrepancy; it was one of the first successes of his new theory of gravity.

Checking the motions of Mercury is a very tricky piece of astronomy. The perihelion of Mercury *seems* to shift by nearly a minute of arc (one-sixtieth of a degree) each year because of the wobble in the Earth's axis. At a rate of about one-tenth of that, the perihelion of Mercury really moves, not because of the einsteinian effect but because of the influence of other planets. The movement explained by Einstein contributes less than one hundredth of the observed movement. In recent times Irwin Shapiro and others have used careful radar observations of Mercury to help confirm the shift due to the einsteinian effect to within half a per cent of the prediction.

An object which shows the required swivelling of the orbit far more clearly than Mercury does is a pulsar, a pulsating neutron star, discovered by radio astronomers using the Arecibo radio telescope in Puerto Rico. Joseph Taylor and his colleagues from the University of Massachusetts came upon this pulsar in 1974; it is a relativist's delight. Its cumbersome name, Pulsar 1913 + 16, refers to its position in the sky. More often it is called the 'binary pulsar' because it is orbiting very closely and rapidly around another collapsed star, so forming a 'binary' or two-star system. It lies a long way off, perhaps 15,000 light-years from us. The companion, so far unseen, may be a white dwarf, a second neutron star or a black hole.

'Flashing' with radio energy about seventeen times a second the pulsar serves as a very precise clock. Its orbit sweeps the pulsar first towards and then away from the Earth, as revealed by the doppler effect, alternately speeding up and slowing down the 'clock' every eight hours. After watching the binary pulsar for several years, Taylor is able to predict its actions with great precision. He can tune in to the pulsar after a lapse of several months and find the first pulse coming in precisely on cue, within less than a thousandth of a second of the predicted moment of arrival. The swivelling of the orbit is conspicuous. The pulsar's position of closest approach to the unseen companion advances by four degrees a year—30,000 times faster than the advance of Mercury's orbit around the Sun. From Einstein's theory one can deduce that the combined masses of the pulsar and its companion are 2·83 times the mass of the Sun.

The radio astronomers have years of work ahead of them, to figure out the masses of the two objects and check all the details against Einstein's theory. One of the most exciting possibilities is that they will see a slow but steady change in the speed of the pulsar in its orbit. The reason is that the rapidly whirling masses in this 'binary' system should be shedding energy by radiating it in a novel form. As a result, the two objects should fall closer together, and speed up. If the change in speed is detected, it will be indirect evidence for the existence of gravity waves. Those form the subject of the next chapter. 'Gravitational radiation' is the correct term for them, because 'gravity waves' is a term adopted by other physicists for more commonplace waves in fluids affected by gravity. But some professors who hunt for waves of the einsteinian sort take the liberty of calling them 'gravity waves', and so shall I.

11 : Waves of Gravity

Warped space stretches objects by a tidal force.
Einstein predicted tidal ripples, or gravity waves.
Gravity waves will stretch objects in alternating directions.
Gravity waves are extremely weak.
Violent cosmic events should create detectable gravity waves.

The tides of the sea are among the most impressive of everyday phenomena. When tidal currents flow among the rocks, sailors are wary. To see a harbour emptying itself of water, leaving boats canted on the mud, and then to think of the water across millions of square miles that will have to heave itself up within a few hours to make them float again, sharpens one's sense of the power of gravity. The tides are a vivid manifestation of the curvature of space in the vicinity of the Sun and the Moon.

As cosmic bodies go, the Moon is but a pebble yet, because it is so close to the Earth, its tidal influence is more than twice as great as the Sun's. So we can simplify the story by considering only the Moon's effect. The tides are due to the *differences* in gravity at different distances from the Moon.

When a sailor in the Atlantic sees the Moon overhead, he is 8000 miles nearer to the Moon than his opposite number in the western Pacific, on the far side of the Earth. The world's oceans are thus exploring different regions of the Moon's sphere of warped spacetime. Compared with the average gravitational effect of the Moon on the Earth, the water on the near side, where the Moon is overhead, experiences a stronger effect. It falls into the timeshells of slower moontime: we see it as an upheaval. On the far side, the Moon's gravity is less and the ocean water rises away from the Moon. In regions where the Moon is seen on the horizon, the sea level is at its lowest.

The overall effect is to distend the sphere of sea level into an eggshaped (ellipsoidal) form, pointing at the Moon. The Moon itself has a frozen shape like that and, as a result, the tidal effect of the Earth on the Moon keeps the same 'end' of the Moon pointing at us always. But the Earth spins, and as each region enters the zone of distension it experiences a rise in sea level. The 'egg' is double-ended, so high tide occurs roughly twice a day. Because friction between moving water and the land makes the rotating Earth drag the ocean bulge around with it, the tides are a little 'out of phase' with the apparent motions of the Moon and the Sun. Moreover, the friction causes the Earth's rotation to slow down. Our 24-hour day is growing longer at a rate of about one hour every 200 million years.

For present purposes, the tides signify curved space. The curvature of space is a measure of the bending of light and is related to the rate at which clocks are changing and gravity is intensifying, as you move closer to its source. But the distension of the ocean water along the line to the Moon is a direct indication of intensifying gravity.

In the curved space near a massive body, space is destroyed: the area and volume of a sphere are less than you would expect from the radius. Accordingly, while an object falling towards the massive body is stretched lengthways by the tides it is also squeezed from the sides to reduce its volume and make it fit into the shrunken space. The force that drives the apple to the ground and guides the Moon in its orbit is an illusion, according to Einstein. But curved space and the tidal forces which it sets up are real. They could tear a man to pieces.

The deformations of the tidal effect are gentle enough in ordinary circumstances, but in the vicinity of a black hole they would become most dramatic. An astronaut falling into a black hole would be stretched and squeezed so strongly that he would become like a long length of spaghetti, even before he reached the surface of the black hole. The egg-shaped oceans and the spaghettification of the astronaut give tangible expression to the local curvature of space.

Another sign of tidal stretching is that, if you drop two stones over a precipice, one after the other, the distance between them will increase as they fall. The leading stone, having begun to accelerate sooner, will always be travelling faster than the second stone. The *relative* acceleration of the two stones is another sign of the curvature of space.

A spaceship with the motors switched off feels no force of gravity— it is weightless—but prudent astronauts of the future may want a system for detecting black holes. They will be able to detect invisible massive bodies by the tidal effect. In one simple method, a flotilla of spaceships travelling in company can monitor the distances between the ships. If they adjust their speeds in empty, uncurved space so that they are flying in perfect formation, any encounter with curved space will disarrange the formation, in a manner detectable by radar. Intership distances along a line pointing towards the black hole will increase; ships spaced along a line at right angles to that line will move closer together. Little by little the flotilla will form itself into an arrow pointing at the black hole.

One of Einstein's most remarkable conclusions was that packets of curved space—tidal ripples, in effect—should travel through empty space, far from the massive objects that created them. Nothing would bring curved space to life better than to sense the curvature changing: to feel a disturbance running through space like an earthquake. That is one reason why the search for 'gravity waves' became an obsession of experimentalists in the late 1970s.

Einstein predicted gravity waves in 1916, as a quick by-product of his theory of gravity, in much the same way as James Clerk Maxwell had earlier predicted electromagnetic waves as a consequence of his unified theory of electricity and magnetism. The parallel goes further. Electromagnetic waves are created by the jerking or vibration of electric charges. In a radio transmitter, electrons oscillate rapidly to and fro; in an atom, electrons can 'jump' into a different orbit, creating visible light in the process; in a hospital X-ray machine, a beam of energetic electrons smashes into a target and the violent arrest of the electrons produces the X-rays. Similarly any vibration or jerking of masses ought to produce gravity waves. And, just as an electromagnetic wave exerts a force at the end of its journey by shaking other electric charges, so a gravity wave can in principle travel through space and shake other masses.

Unlike most forms of energy, a gravity wave can also pass right through solid objects—the Earth for instance—with very little weakening. Its effect is like an extra 'dose' of tide, repeatedly changing its orientation as it passes. If you imagine a gravity wave coming straight down on the Earth through a football field, it will first make the field slightly longer but narrower, and then slightly shorter but wider.

Gravity waves are a form of communication in the universe. In fact a very simple argument makes them seem indispensable. Stars and planets move about, and violent events alter the arrangement of matter in space. The question arises: how do objects 'know' that the pattern of gravity is changing too? Suppose, for example, that the Moon fell into the Earth. There are Earth satellites, lunar satellites and natural debris in the Earth-Moon system, whose motions under gravity are governed by the existing positions. They cannot continue in their old orbits so they must be notified of the change by signals. That implies a flow of energy, something like gravity waves.

But the gravity waves are expected to be extremely feeble. This is a matter of some torment for the experimenter who sets out to detect them. Einstein himself reckoned the energy radiated from the moving arms of an object like a windmill. It turned out that the windmill, steadily rotating for a million years, would convert only about one billion-billion-billionth part of its energy of rotation into gravity waves. The weakness of gravity waves is congenial for life on Earth. If our planet had shed much energy into gravity waves it might have crashed into the Sun long before the human species evolved.

The chief reason for this feebleness is that the centre of gravity does not move: any shift of a mass in one direction is inevitably balanced by the shifting of another mass in the opposite direction, and the gravity waves tend to cancel out too. The only thing that would change if the Moon fell into the Earth would be the precise arrangement of the masses. This would make very little difference to, say, the gravity of the Earth-Moon system felt by the planet Mars.

You have to invoke very powerful sources indeed to produce gravity waves with even the slightest hope of their being detected on Earth. Stars which are orbiting closely around each other in a 'binary' system should give off fairly strong gravity waves continuously, especially if they are very close together. But even if you added together all the gravity waves arriving at the Earth from all the binary stars in the Milky Way Galaxy, the resulting 'buzz' would still be extremely difficult to detect. The convulsions accompanying the explosion of a giant star should give off far more powerful gravity waves: that is probably one of the better bets for a source detectable with present instruments. The snag is that such events are rare—about one every hundred years in our Galaxy. Very massive 'star-swallowing' black holes at the hearts of other galaxies should generate more frequent gravity waves. Collisions between black holes would be particularly energetic sources, and perhaps not as rare in the universe as one might suppose.

The first builder of gravity-wave detectors was Joseph Weber at the University of Maryland. In the mid-1960s, his major instrument consisted of a cylinder of aluminium weighing 1 1/2 tons, intended to vibrate if a gravity wave of an appropriate frequency should pass across it. Hanging from vibration-proof mountings in a vacuum chamber, the cylinder carried a girdle of piezo-electric crystals, which responded to the slightest stress by producing an electric voltage. Two such 'antennas' for gravity waves operated together, one at Weber's university and the other at the Argonne National Laboratory, 600 miles away near Chicago. They were later supplemented by an aluminium disc in Maryland.

Weber's pioneering labours were magnificent but, in the opinion of many physicists, he claimed the detection of gravity waves prematurely. Agitation due to the heat energy in the cylinders and the electronic equipment produced a lot of false signals in the records of the antennas, both at the University of Maryland and at Argonne, but about once a day the cylinders, far apart, would seem to shake at the same moment. Weber said that only about one in six of these coincidences was due to chance and that the rest were probably caused by gravity waves. He suggested that they were coming mainly from sources close to the centre of the Milky Way Galaxy.

The trouble was that Weber appeared to be seeing far too many bursts of gravitational radiation. To account for them one would have to assume the annihilation of matter equivalent to at least several Suns every year, in the heart of the Milky Way. The most generous-minded astrophysicists were unable to say how so much destruction could be going on without any clear outward signs other than the trembling of the cylinders. Weber himself recognised the 'energy problem' as he called it, and he commented: 'The very large energy release that is implied is the strongest reason to suspect that the results of the experiments are not completely understood.' In any case, when other experimenters in the USA and Europe built detectors of similar sensitivity they simply could not register or confirm any of the gravity waves that Weber said were so copious.

By the late 1970s, as the Einstein centenary approached, new gravity-wave detectors were being prepared at a dozen laboratories scattered around the world. They were of three main kinds. The first were direct descendants of the Weber cylinders, but now intended to be about a million times more sensitive. The chief step forward was to operate them very close to the absolute zero of temperature, to get rid of false signals due to heat energy that so plagued Weber. Their construction, a monumental exercise in engineering, was undertaken at the Argonne laboratory (with Weber participating), at Stanford University in California, at Louisiana State University, at the University of Rome and at the University of Western Australia in Perth.

As an example, the Stanford instrument developed by William Fairbank and his colleagues uses a five-ton cylinder of aluminium. For sensing small motions due to the putative gravity waves, the Stanford experimenters have novel low-temperature detectors called 'squid' sensors. An extraordinary system of insulation and refrigeration is designed to cool a huge mass to within a tenth of a degree of absolute cold ($-273°$ Celsius).

The second notable kind of ultra-sensitive detector uses the most perfect jewels in the world: man-made sapphires grown as large single crystals. The idea came from Igor Braginsky in Moscow, an outstanding experimentalist of relativity. A single crystal is a very effective 'bell', so much so that it will tend to vibrate continuously because of internal heat, even at very low temperatures. This turns the spurious signals into a 'hum', against which the 'clink' of a gravity wave should seem quite distinctive. Detectors like this have been under development at the University of Moscow and at the University of Rochester in the USA.

In the third main class of detectors of the new generation the experimenters avoid heroic low-temperature engineering and instead seek higher sensitivity with very delicate systems, involving laser beams, that look for relative movements between widely spaced masses. (The principle is similar to that recommended a few pages back for black-hole detection by the tidal effect on a flotilla of spaceships.) The masses are arranged in a right-angled triangle or a square so that a gravity wave going through will tend first to lengthen one 'arm' of the right angle while shortening the other, and then reverse its treatment of the arms. The 'separated-mass' detector originated with the Hughes Research Laboratory in the USA, and matured among other groups, notably at the University of Glasgow, the Massachusetts Institute of Technology and the Max-Planck Institute for Physics in Munich.

In a detector built by Ronald Drever and his colleagues at Glasgow in 1978, four masses are arranged in a square about thirty-three feet apart. A laser beam reflected many times between them serves to detect changes in separation between them as small as a million-billionth of an inch—far smaller even than the nucleus of an atom! But even this extraordinary precision may be inadequate to detect any but very rare, violent outbursts of gravity waves. The physicists therefore bend their minds to devising other systems—to operate in space, for instance, or even to circumvent the 'uncertainty principle' which, at first sight, limits the possible precision.

The 'binary pulsar', described in the previous chapter, should be losing energy by the emission of gravity waves. If this effect conforms with Einstein's theory it should be apparent in the motions of the pulsar by the early 1980s. Observations of the binary pulsar may prove the existence of gravity waves indirectly, before the detectors on the Earth begin to show direct evidence for them.

The bare detection of gravity waves cannot be the end of this story. One or two good experiments may suffice to establish the waves' existence and confirm Einstein's prediction in its baldest form, but physicists will want to examine details, especially the speed of the waves. The simplest way to check this will be to spot a visible event such as the supernova explosion of a star and to detect gravity waves from the same event. If Einstein was right, the burst of gravity waves will arrive at the Earth at the same time as the light flash, because they travel at exactly the speed of light. The form of the waves and its resemblance to the predictions will be another important test of General Relativity. Rival theories predict gravity waves subtly different in character. Some of them, advertising more intense gravity waves than Einstein's, are already ruled out by the observations of the binary pulsar, which is not losing energy quickly enough to satisfy those theories.

The deployment of a number of gravity-wave detectors around the world or in space will make it possible to establish the direction in the sky from which the waves are coming. 'Gravity-wave astronomy' will then have a wide-open future as a new method of looking out on Einstein's universe. It should dovetail nicely with the detection of supermassive black holes in exploding galaxies and quasars, which will almost certainly be the most reliable sources of gravity waves, as ordinary matter falls into them. And rare but stupendous 'spacequakes' produced by collisions between black holes may be detectable at distances of billions of light-years.

A postscript on gravity waves. Just as electromagnetic waves consist (as Einstein discovered) of particles of light, or photons, so gravity waves are said to consist of gravitons. Describing the actions of these particles is another route to understanding the curvature of space. Gravitons supposedly communicate the effects of gravity between all objects in the universe that possess mass. But gravitons themselves possess energy and are therefore heavy, so they themselves are vulnerable to the action of other gravitons. In other words the particles conveying the effects of gravity feel the effects themselves—and they are therefore deflected along curved tracks. This incest among the gravitons produces the curvature of space.

Albert Einstein at the height of his powers in 1916, when he had just completed his theory of relativity. His fellow-physicists were perplexed by his ideas, but Einstein awaited the verdict of nature with complete confidence. (*Photo: Radio Times Hulton Picture Library*)

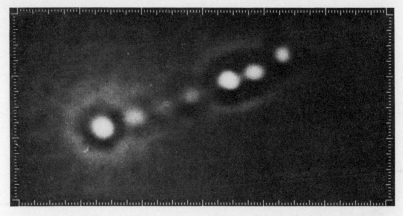

Six knots of luminous gas are visible in the jet expelled from the stormy galaxy M87. The large single blob is the centre of the galaxy, which is said to contain a supermassive black hole. The dark haloes are artifacts of the special image-processing technique used by Halton Arp and Jean Lorre to generate this unusually clear picture of M87's jet. *(Photo: Halton Arp)*

A primitive particle accelerator, built by John Cockcroft and his Irish colleague Ernest Walton in Cambridge in the early 1930s, produced nuclear transformations that verified $E = mc^2$, Einstein's formula for the equivalence of mass and energy. *(Photo: Cavendish Laboratory)*

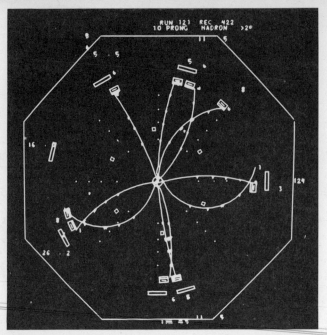

The creation of matter is recorded at the Stanford Linear Accelerator Center. From the head-on collision of an electron and an anti-electron there emerges a swarm of newly created particles that are intrinsically far heavier. Energy of motion is converted into the rest-energy of matter in accordance with $E = mc^2$. *(Photo: Stanford Linear Accelerator Center)*

At the heart of the Crab Nebula is a rapidly pulsating 'star', or pulsar, flashing thirty times a second. As remnants of a massive star that was seen to explode in AD 1054, the nebula is debris hurtling into space, and the pulsar is the collapsed core of the star. A similar explosion could create a black hole instead of a pulsar. *(Photo: Hale Observatories)*

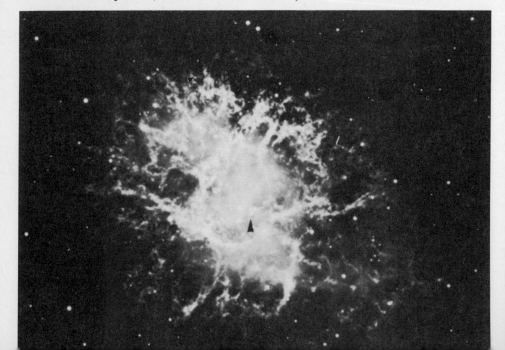

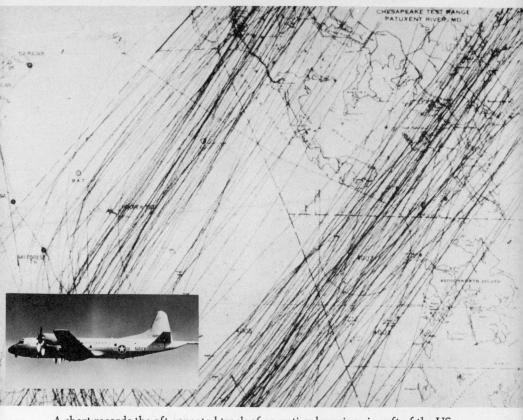

A chart records the oft-repeated track of an anti-submarine aircraft of the US Navy *(inset)* flying over Chesapeake Bay to check Einstein's predictions about the effects of gravity on clocks. *Below* is an airborne box enclosing atomic clocks in a carefully stabilised environment. Five flights confirmed that clocks run faster at high altitude, where the Earth's gravity is weaker. *(Photos: Carroll Alley)*

The most perfect ball ever made is checked at Stanford University in prepara-
tion for a subtle test of Einstein's General Relativity in the 1980s. Four such
balls, made of fused quartz, are to be coated with a superconducting film and
set spinning as gyroscopes in a spacecraft orbiting the Earth. The experiment
is designed to detect the very slight dragging of space by the rotation of the
Earth. *(Photo: Francis Everitt)*

A ball-bearing rolled at an appropriate speed on a suitably deformed surface
will orbit like a planet going around the Sun. Curved space is often represented in this fashion using a rubber sheet, but the photograph is a close-up
of a hardened version built by BBC-TV in the form of a distorted 'pool table'.
The large ball represents the Sun. *(Photo: Joan Williams)*

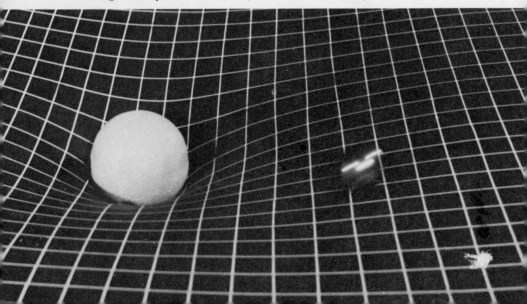

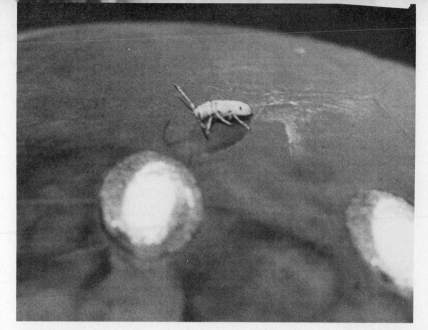

A bug crawling on a balloon decorated with galaxies mimics a space traveller in a 'finite but unbounded' universe. It might march in what it judges to be a straight line, and yet finish up back at its starting-point having unwittingly circumnavigated the universe. *(Photo: Joan Williams)*

When the Moon totally eclipses the Sun, stars become visible beyond the Sun and, in Einstein's theory, their apparent positions in relation to other stars are slightly shifted by the effect of the Sun's gravity. In this negative photograph, prepared by the Royal Greenwich Observatory from one of the historic plates of the 1919 expeditions, the Moon's dark disc appears white, and the Sun's bright corona is dark and fuzzy. Two stars are marked—black dots not easily distinguished from flaws and dirt. *(Photo: Royal Greenwich Observatory)*

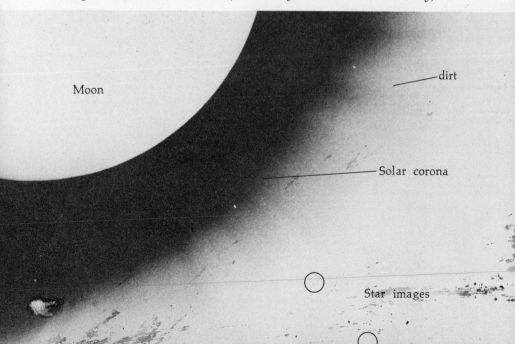

Moon

dirt

Solar corona

Star images

The Haystack radio telescope in Massachusetts has
served in important tests of General Relativity,
notably in radar-ranging to the planets. The 120-foot
dish is housed inside a protective dome, which is
transparent to radio waves.
(Photo: Lincoln Laboratory)

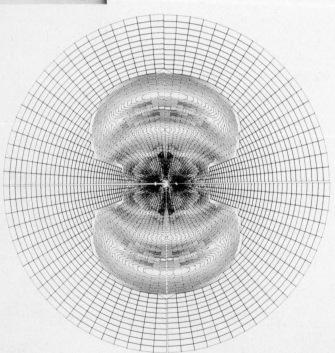

Waves of gravity pour from an extraordinary (and imaginary) cosmic accident
—a head-on collision of two black holes approaching from left and right.
Kenneth Eppley and Larry Smarr used a computer to calculate and draw con-
tours of the variable curvature of space that constitutes the radiation. Such a
collision would convert about a thousandth part of the rest-energy of the
black holes into gravity waves. *(Photo: L. Smarr and K. Eppley)*

Gravity-wave detectors under development at the University of Glasgow use laser beams to measure very small changes in the separation of two objects—two metal cylinders lying end-to-end, in this early version. A system of mirrors and lenses can be seen at the side of the cylinders; it guides the laser beam so that it makes many reflections, thereby increasing the sensitivity in measuring slight movements. *(Photo: Ronald Drever)*

The equipment deployed on the Moon by the *Apollo 11* astronauts includes a slanting box, seen in the middle distance. This 'laser-ranging retroreflector' contains corner reflectors, like that shown bottom-out in the *inset,* and redirects laser light from the Earth along a return path. *(Photos: NASA; Perkin-Elmer [inset])*

At the McDonald Observatory of the University of Texas, a laser is aligned for firing pulses of light at the Moon, using the observatory's 107-inch telescope. The observer aims the telescope at one of the man-made reflectors on the Moon, and the electronic equipment at the far end of the room times the returning light particles to better than a billionth of a second. Ten laboratories were jointly involved in setting up this experiment, which provides one of the most delicate tests of Einstein's theory of gravity. *(Photo: Joan Williams)*

Life extension for particles, equivalent to stretching a human life to 2000 years, was accomplished in a ring of electromagnets at the CERN laboratory near Geneva. Unstable "muons" circulating at close to the speed of light survived nearly thirty times longer than similar particles at rest. *(Photo: CERN)*

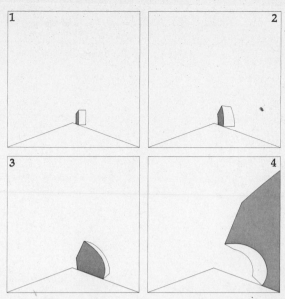

A building (1) would appear deformed (2) to a traveller passing down the street at a substantial fraction of the speed of light. The building twists away from him (3), showing less of the near wall and more of the side. The traveller then begins to see around the corner, to the far wall of the building (4). If, for example, the traveller is going at half the speed of light, he begins to see around the corner when the corner lies 60 degrees off his track. For faster speeds it occurs earlier, when the building is still a long way off. (Drawings adapted from the computer-generated film *Cubes*, made by Edwin F. Taylor.)

The Small Magellanic Cloud, one of the nearest galaxies, contains a pulsating source of X-rays, SMC X-1. After studying the pulses crossing vast distances from this and other X-ray pulsars, Kenneth Brecher declared that the speed of light varied by less than the speed of a turtle. (*Photo: Royal Observatory, Edinburgh*)

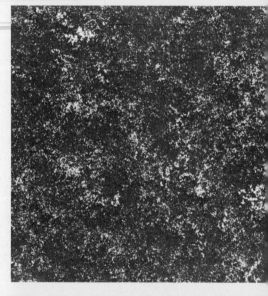

Computer simulations show, *top*, a
wholly random scattering of galaxies
and, *centre*, the clustering due to
mutual attraction between the
galaxies. The latter picture resembles
far better a real map of galaxies,
below. From such studies James
Peebles and his colleagues at
Princeton University have concluded
that the density of matter in the
universe is greater than previously
supposed. *(Photos: James Peebles)*

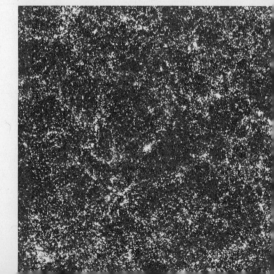

Albert Einstein and the Danish theorist Niels Bohr admired each other greatly
but they were implacably opposed in their opinions about modern quantum
theory, which Einstein assailed and Bohr defended. From their great debate
in the late 1920s, Bohr emerged the victor. *(Photo: Paul Ehrenfest, Courtesy
Einstein Trustees)*

12 : The Galileo Mystery

The effects of gravity and acceleration are equivalent.
Everything must fall at the same rate.
Light bends in an accelerating spaceship.
An accelerating spaceship could create a black hole behind it.
The Moon's motion confirms Einstein's 'equivalence principle'.

Falling is the most natural way for objects to behave and our main mode of travel through the universe, aboard our planet. Nevertheless in relation to the Earth itself we spend most of our time not falling. There is not really a black hole at the centre of the Earth, although apples and landslides and the man who slips on a banana skin try to move towards the centre just as if it were there. They find firm ground barring their way. The atoms of our planet are comfortingly strong compared with the effect of gravity and they build rigid rocks for us to stand on. 'Not-falling' appears in Einstein's theory of gravity as an acceleration —a continual gathering of speed—even when you are staying in one place.

A helicopter offers a clue to this interpretation. It has to keep its engine working frantically to avoid falling. The pilot and his passengers feel their seats pushing upwards even when the helicopter is plainly just hovering at one place. But they are perfectly comfortable, because that is exactly what they feel if they are sitting on the ground. If the helicopter now climbs quickly, the seats push harder on the people. They seem heavier and they feel their stomachs tending to be left behind. The push upwards is often called a 'g-force'. In free fall it is zero 'g'. On the Earth's surface or in the hovering helicopter it is one g. In a spaceship being blasted into space it is several g and g-forces are, for a while, very strong and unpleasant.

Now imagine an astronaut in a very fast spaceship, coming home after a long trip but with plenty of power to spare. He has to lose a lot of speed before he reaches the Earth, which means that he must swivel his ship around and fire his rockets forward, for a long time, to act as a brake. To be comfortable, he adjusts the rockets so that the g-force is one g. He then stands or sits with his feet pointing in the direction of travel and feels exactly the same as he would feel on the Earth. Satisfied that his computer has the flight-plan perfectly set, he puts shutters over all windows so that there is no risk of them melting in the friction of the Earth's atmosphere.

His friends and relations come to the spaceport to meet the astronaut. They see the ship come smoothly down, losing speed all the time. It approaches to within six inches of the ground and stops. Its motors are still blasting away, so the friends and relations dare not approach. They wait for the astronaut to switch off the motors and come out. But the astronaut is reading a book. With his windows sealed he is completely unaware that he has arrived home. He knocks a tumbler off his table and it falls, hitting the floor exactly as it would on the Earth—or exactly as is to be expected in space when the ship is accelerating or braking at one g. Meanwhile he is still hovering six inches above the ground, wasting enormous amounts of fuel, polluting the atmosphere and keeping the welcomers waiting.

They might wait for weeks. Einstein realised that there is no simple way of distinguishing between being on the surface of the Earth and being in a spaceship with the motors running at the appropriate rate. On the contrary, everything matches. If the astronaut tosses a ball into the air in his cabin it will go up and come down again just as you or I would expect. In the spaceship, every object weighs exactly the same as it would do on the Earth.

Einstein's interpretation of the appearances of gravity on the Earth as being almost indistinguishable from acceleration through space cleared up the Galileo Mystery at a stroke. This was the question of why (in a vacuum at least) everything falls at the same rate, accelerating exactly in company. In Einstein's relativistic view you can think that it is the ground which is accelerating upwards, as if powered by a million rockets, while the falling objects are at rest. If that is the concept there is no surprise whatever that the ground approaches all of the falling objects at exactly the same rate of acceleration.

Formerly there were two distinct concepts of the mass of an object: one expressing itself under gravity as weight and the other showing itself during acceleration as resistance to acceleration, or 'inertia'. Before Einstein, it was puzzling that the same figure would describe both exactly. By interpreting gravity as an acceleration and clearing up the Galileo Mystery, Einstein reduced a double problem to a single one. But the fact remains that weightless objects in a freely falling spaceship suddenly become heavy and resist acceleration, as soon as the motors are switched on.

There is no surprise that the floor of the spaceship comes up and hits them; what needs explaining is why they resist vigorously and the spaceship has to consume power in pushing them forward. It is not a matter of speed—when the motors stop the objects all become weightless again, even if they are now going twice as fast as before. Mass takes effect only during acceleration, or its gravitational equivalent. And here a third meaning of mass comes into the picture: Einstein's identification of mass with energy. The more energy or mass that an object possesses the harder it is to set moving, and the extra energy needed to accelerate it to a given speed depends upon the energy it already possesses.

Reverting to the Earth: the ground is plainly not being pushed upwards by a million rockets. Apart from any other consideration, if the ground were really accelerating non-stop at thirty-two feet per second per second it would approach the speed of light within a year. In any case the idea is a gross affront to common sense, which is my excuse for postponing such a crucial and early point in Einstein's thinking until this late stage in the explanation of his theory of gravity. The correspondence between accelerating in a spaceship and standing on the Earth can be appreciated only in the context of the various effects of distorted space and time in the vicinity of the Earth which create the appearances of acceleration. But gravity is an older art than spaceflight, so it may be better to invert the argument and say that an accelerating spaceship creates an illusion of gravity.

The correspondences are indeed amazing, if you liken gravity to acceleration. For example, it follows at once that light yields to gravity, and this is what I hinted at when I mentioned that Einstein had a stronger reason than the mass of light for expecting the effect. If the astronaut aims a beam of light sideways across his spaceship cabin, it will hit the wall on the far side slightly lower down than it would if the spacecraft were not accelerating. The spaceship tends to leave the particles of light behind and, in relation to the spaceship, the light is curving somewhat downwards towards the floor, as if under gravity.

More subtle equivalences appear. Einstein's gravitational redshift—the reduced energy of light coming from a massive object—translates exactly into Doppler's redshift due to the relative motion of the source of light and the observer. If the astronaut looks down at a footlight on the floor of his accelerating spaceship he will see the light slightly reddened because, in the interval between the light leaving the lamp and arriving at his eye, the spaceship and the astronaut have gained a little more speed.

In this special sense, the lamp and the eye are moving apart. The effects on clocks, too, are exactly the same as under gravity. An atomic clock on the ceiling of the accelerating spaceship will be seen to run slightly faster than a clock mounted on the floor. Look up, and the acceleration carries you towards the clock, so that you see the clock registering its next second a little sooner than you would if the spaceship were travelling at a steady speed; look down, and the indications of the other clock are delayed, so it is running slow.

Implied in this description of the astronaut's view of his clocks is the fact that light does not appear to travel at the same speed up from the floor, as it does from the ceiling. A cardinal rule of relativity, that light always seems to travel at the same speed, applies only to systems that are moving at a steady rate in empty space, or else falling freely under gravity. Just as gravity affects the path and speed of light and the rate of time, as judged by a distant observer, so does acceleration in a spaceship. If you are *accelerating* towards a source of light, its speed seems greater. If you are accelerating away from it, its speed seems diminished and during prolonged acceleration the light, coming from behind you, will have difficulty in catching up with you.

The real spaceships of the twentieth century accelerate only for short periods and then coast in free fall. But if you imagine a long-playing nuclear propulsion system that could sustain a steady acceleration indefinitely, another fascinating imitation of gravity appears. The spaceship creates a black hole in its wake. To understand the high-speed traveller's equivalent of a black hole, it helps to recall the old Greek paradox about the race between Achilles and the tortoise which has a head start.

The athletic hero runs to the tortoise's starting point, but meanwhile the tortoise has ambled a short distance ahead; Achilles covers that extra distance, only to find that the tortoise has inched ahead again—and so on *ad infinitum*. Achilles, it seems, can never overtake the tortoise, which is why he sulks in his tent. A charming story, a plausible argument, but nonsense of course. The solution to the puzzle is that an infinite series of events, like that described, can in fact be completed in a moment.

Now suppose that the tortoise is a supercharged animal that makes the usual slow start but *accelerates.* Then the paradox reasserts itself in a novel form because the tortoise can elude Achilles even if it never runs quite as fast as Achilles does. The reason is that, although Achilles can keep narrowing the distance, the time required for him to complete the task and overtake the animal keeps growing longer and longer. For example, by the time Supertortoise is going at 99.9 per cent of his speed, Achilles' overtaking time from one hundred yards behind is already several hours; a little more speed from Supertortoise and it becomes years, then centuries. Unlike the original paradox, this one is not false. The fact is, you can elude a pursuer for ever even if you never run exactly as fast as he does.

For Supertortoise read astronaut; for Achilles read a ray of light coming up from behind him. The astronaut cannot surpass the speed of light but by accelerating continuously he can (in principle) get closer and closer to it, just as the tortoise approached the speed of Achilles. And in those circumstances the beam of light will never catch up with the astronaut.

If the accelerating astronaut looks out of his rear window he can see nothing: the stars have vanished in that direction, because their light cannot reach him. As far as he is concerned, a segment of the universe has become a black hole. He can achieve this effect (again in principle) without any arduous sensations; provided his spacecraft is powerful enough and well shielded in front, a comfortable acceleration of one g will achieve the trick within a year. Moreover, as Paul Davies of London and William Unruh of Vancouver have realised, the accelerating spaceship is slightly warmed. Just as a gravitational black hole squeezes radiation out of empty space, so does an accelerational black hole.

The equivalence of acceleration and the experience of gravity is the cosmic principle of einsteinian physics. There are two versions of it. The 'weak' equivalence principle says, as Galileo did, that all objects fall at the same rate under gravity. The 'strong' equivalence principle declares that the laws of physics are the same everywhere and at all times, throughout the observable universe, despite any effects of motion or gravity. The latter is what Einstein craved for during his years of mental struggle. But it is demanding quite a lot because the laws of physics embrace, besides motion and gravity, the laws of electricity, atomic physics, heat and the behaviour of every form of matter. 'Everywhere and at all times' spans billions of light-years of space and billions of years of time. Experimenters have been at considerable pains to check these propositions.

Because the electric and sub-atomic forces are stronger than gravity, it is easier to check them for any discrepancies. Astronomers see atomic light in the most distant galaxies and quasars behaving just as it does in laboratories on the Earth. Geology reveals no change whatever in the laws of nature during the Earth's history. In particular, the behaviour of particular elements in a natural nuclear reactor operating at Oklo in Gabon two billion years ago shows that these non-gravitational forces have changed by less than one part in a billion over the immense span of time. For them, there is persuasive evidence that the 'strong' equivalence principle is correct. In the case of gravity itself, the problem is trickier.

The first requirement is that Galileo's law should be not just roughly right (which it obviously is) but exactly right. In the most searching tests imaginable, 'gravitational mass' and 'inertial mass' must be identical.

Between 1886 and 1922 in Budapest, Roland Von Eötvös compared
them very accurately. For instance, he examined the direction of the
vertical shown by plumb-lines using bobs made of different materials,
such as wood, tallow, asbestos and various metals, alloys and salts. At
middle latitudes—at Budapest for example—a plumb-bob feels not
only the Earth's simple gravity but the Earth's rotation as well. The
rotation deflects the plumb-line through an angle of about one hundredth of a degree from the 'true' vertical pointing at the centre of the
Earth. Conventionally, the centrifugal deflection depends on 'inertial'
mass while the effect of gravity depends on the 'gravitational' mass. If
there were any discrepancy between them and gravity acted differently
on different materials, Eötvös' various plumb-bobs would be out of
alignment. He checked that they were not, to an accuracy of about a
billionth of a degree.

In another classic gravitational experiment, Robert Dicke and his
colleagues examined the effect of the Sun's gravity on gold and
aluminium cylinders in the laboratory at Princeton University. Their
set-up would have amazed Galileo in its complexity and delicacy: it
used beams of light and electrical methods for detecting, suppressing
and measuring very slight movements. The rig stood on bedrock, temperatures were controlled to one ten-thousandth of a degree, and even
the experimenters had to keep well away so that the gravity of their own
bodies should not influence the cylinders. But in essence all they were
doing was checking that gold and aluminium fell towards the Sun at the
same rate, just as Galileo had checked that gold and copper fell towards
the Earth at the same rate.

At the Earth's surface, the Earth's gravity is 1660 times stronger than
the Sun's gravity, but because the Earth is spinning the direction of
acceleration due to the Sun keeps changing, being eastwards in the
morning and westwards in the evening. Dicke and his colleagues hung
cylinders of gold and aluminium from a horizontal frame supported on
a fibre. If there were any difference in the effect of the Sun's gravity on
the two materials, the frame would tend to swing slightly, every
twenty-four hours. In 1964 the experimenters were able to show that
the effect on gold and aluminium was the same to within ten parts in
a million million. Some years after that, Igor Braginsky in Moscow
carried out a similar experiment with platinum and aluminium. He
improved the accuracy further, and still no difference was detectable.

Those were very impressive tests of the 'weak' equivalence principle,
but there was a drawback. The masses tested were small and the possibility remained that heavy objects might behave differently. Indeed,
unless the 'strong' equivalence principle is strictly correct, the possibil-

ity remains that gravity deals differently with itself—with the gravitational energy that binds the Earth together, for example—than with other manifestations of energy. Checking that point prompted one of the most interesting experiments in General Relativity.

A problem about testing the 'strong' equivalence principle, that the laws of physics should be always and everywhere the same, is knowing what to look for. But since Einstein wrote out his General Relativity other people have offered other theories which violate the principle. Even for faithful einsteinian relativists who expect that the alternative theories will turn out inevitably to be wrong, they are valuable as foils against which to test General Relativity. To rehearse every possible variant of modern theories of gravity would be tedious—there are dozens of them—so I shall mention just one of the strongest contenders. It is a theory advanced by Carl Brans and Robert Dicke in 1961, although it was anticipated in a general way by Pascual Jordan in 1959. For brevity, I shall call it Dicke's theory. It has tended to lose its contests with Einstein's theory, but this should not be taken as any disparagement of Dicke's work: on the contrary he is one of the foremost experimenters and theorists of relativity.

Dicke's theory of gravity is essentially Newton's theory translated into Einstein's curved space. It disagrees with Einstein's theory in declaring that gravity grows weaker as the universe expands, diminishing by a few per cent every billion years, so that the laws of physics were different a billion years ago and the 'strong' equivalence principle is broken. Dicke's theory offers different values from Einstein's for the expected magnitude of various effects of gravity. The slowing down of radar pulses near the Sun, as measured by Irwin Shapiro, should be a few per cent less by Dicke's theory than by Einstein's; so should the deflection of starlight by the Sun. The orbit of Mercury should swivel a little more slowly. By all these tests, Einstein's theory looks safer than Dicke's.

Furthermore, gravitational energy behaves differently from other forms of energy, according to Dicke's theory. In consequence the Moon should follow a slightly different path in its orbit around the Earth and fall faster towards the Sun than the Earth does—in a subtle violation of Galileo's ideas as well as Einstein's. If Einstein were wrong and Dicke right, the Moon's orbit should be deflected by a distance of several feet towards the Sun. Because the relative directions of Sun and Moon change during each month, small but systematic variations in the distance of the Moon from the Earth would result.

The laser makes it possible to measure distances across a quarter of a million miles of space with the precision needed to look for such very small variations. Partly at Dicke's instigation the necessary experiment was put in hand. In 1965 Russian astronomers had used a 100-inch telescope to direct laser flashes at the Moon and to look for the flashes returning 2.6 seconds later; although the returns were extremely weak the astronomers were able to measure the Moon's distance to within about 600 feet. The Americans improved the laser-ranging method.

In 1969 'Buzz' Aldrin, one of the first astronauts to walk on the Moon, set up a special reflector at Tranquility Base: a panel consisting of a hundred 'cat's eyes' or 'corner reflectors', designed to redirect the laser light as accurately as possible back to an observatory on Earth. Later *Apollo* missions and two unmanned Russian landers put similar reflectors at other points on the lunar surface. With the aid of the reflectors it became possible to measure changes in the distance of the Moon to within a few inches. A group of astronomers and physicists then set out to check the law of gravity with unprecedented accuracy. Since 1969 the lunar laser-ranging experiment has used the 107-inch telescope at the McDonald Observatory in Texas for the most searching test so far of Einstein's theory. Setting up the experiment involved the efforts of about ten collaborating laboratories.

The ordinary astronomical observations at the McDonald Observatory are frequently interrupted when the telescope swings towards the Moon for a 45-minute 'run' of laser ranging. It is fascinating to watch, because the observer must continuously aim the big telescope and the emergent laser beam precisely at one of the retroreflecting panels—and the target is moving. 'It's a sporting occasion', says Eric Silverberg, who oversees these operations at McDonald.

Every three seconds the laser fires a very brief pulse of light, only three feet long. At about the expected moment of return the detection system is activated and it admits a single particle of light. It may be stray light, but an atomic clock times its arrival to better than a billionth of a second and a computer compares the timings for successive particles of light. (Carroll Alley of the University of Maryland, instigator of the flying clocks experiment described earlier, devised the all-important timing system.) When an incoming particle of light arrives at the appropriate instant it is judged to be from the reflector on the Moon, and a bell rings to let the observer know that he is on target.

The bell rings for Einstein. General Relativity emerges completely unharmed from this experiment and there is no trace of the sunward displacement predicted by Dicke. Interpreting the results of the experiments calls for complicated computations, because the distance of the Moon varies overall by 30,000 miles and there are many 'normal' effects of lunar motions much greater than the suspected displacement. After the first six years' work and 1500 good laser echoes from the Moon, the observers and analysts were able to affirm that the Moon's motions did not depart from einsteinian expectations by so much as one foot. If Dicke's theory survives at all, it can do so only in a much diluted form that is virtually indistinguishable from General Relativity.

The laser ranging continues at McDonald and one reason is to see whether the Moon's motions reveal any evidence for a change in gravity with time. Dicke's is not the only theory in which gravity is said to grow weaker with time and Paul Dirac, who predicted anti-matter, offered in 1937 a theory with that same ingredient. In the mid-1970s Thomas Van Flandern of the US Naval Observatory claimed that he had detected a progressive weakening of gravity in an increase in the length of the month—the period of the Moon's orbit around the Earth.

The month grows longer, whether gravity is weakening or not, because the Moon steadily gains energy from the Earth by an effect of the tides. It therefore moves farther from the Earth and adjusts its speed to travel more slowly. Because of this mechanism each month should last a few millionths of a second longer than its predecessor. But after studying the Moon's motions over twenty years, as timed with atomic clocks, Van Flandern came to the conclusion that the month was growing longer at almost twice the expected rate. He offered the weakening of gravity as an explanation.

If further studies were to confirm that interpretation it would damage Einstein's theory, which has no room in it for any such change. But at the time of writing there is more than a suspicion that some of the early atomic-clock readings were not very reliable, and that newer data may fail to show the effect reported by Van Flandern. Most relativists are satisfied that radar-ranging to the planets would detect any significant change of this kind in the intensity of gravity, and has not done so. Radar-ranging will confirm such changes in the intensity of gravity, if they are real, but most theorists expect that this challenge to General Relativity will falter, as others have done. As Roger Penrose of Oxford was able to declare in 1978: 'General Relativity is at least very close to the truth.'

13 : Methuselah in a Spaceship

Biological time matches atomic time.
Visiting a black hole could keep you young.
High-speed travel could keep you young.
Special Relativity deals with high-speed motion.
Unstable particles 'live' longer when they travel rapidly.

Are you ready to grasp firmly the nettle of einsteinian time? The more thoroughly the theory of General Relativity is borne out by experiments, the less avoidable its implications become. So long as one is talking of a microsecond here or there, relativistic effects on time may seem to be of no great consequence. But the modern theory of gravity makes it quite clear that significant time-travel into the future is possible in principle. The idea that time stands still at the edge of a black hole has to be taken quite literally. A black hole could be used to stretch a person's life and let him survive for thousands of years into the future.

It would be a one-way journey. This is not time-travel of the kind described in science fiction, where you can visit the past and interfere with history, or go off into the future and then return to tell your friends what it will be like. Passing back and forth through time like that entails horrendous logical problems in the workings of the universe, which Einstein himself would abhor. But no contradictions arise when relativity allows events to unfold more slowly for one observer than for another, including the events of a person's living, ageing and dying. The moment has come to affirm that everything I have said about changes in the rates of atomic clocks applies in full measure to the rates of life itself.

The operations of atomic clocks show that atoms certainly run slowly under strong gravity, compared with the rate at which they run in empty space. The colours of light emitted by atoms in natural circumstances testify to the same effect. But the human body is a collection of atoms reacting endlessly together in the elaborate molecular dance of life. The rates of all the essential living processes are governed by the rates of atomic action. Under strong gravity brain impulses, for example, will pass less rapidly and hearts will beat more slowly. Einstein would rightly regard this as needless elaboration. For him all energy diminishes, or slows down, in the presence of gravity. Anything that feels the effects of gravity is subject also to the effects on time.

When the propositions about time in General Relativity are interpreted at face value, the practical consequences on the Earth are utterly trivial compared with the philosophical shock induced in pre-einsteinian man, by the realisation that the recording of time is a private or parochial matter. Even when all the arguments have been rehearsed, the mind looks for an escape, back to absolute time. If theory and experiment agree that clocks run more slowly 'there' than they do 'here', some people still suspect that it is an illusion—that real time is 'here' and familiar and anyone I know who goes 'there' will in a certain sense always be adding the same number of minutes and years to his age as I do. The conclusion from relativity is that by going 'there' a person can in principle slow down his clock, his calendar and his life-rate and so outlive those who remain 'here', travelling effortlessly into their future. He will experience time passing at a rate that seems quite normal to him, and yet is much slower than 'here'.

A more promising escape route might be to suppose that there remains some absolute cosmic time, marching steadily on, which different observers—including ourselves—measure falsely and differently because our clocks are less than perfect. For example, if you took an atomic clock far enough away into empty space, it would cease to feel the gravity of the Earth or the Sun and might then be said to be approximate to the 'true' cosmic rate. While there may be some practical sense in such a notion, it runs counter to the spirit of Einstein's theory, which takes the democratic view that everyone's assessment of events is equally valid. If a time-traveller looks out and sees the universe apparently speeded up, because time for him is running slowly compared with time in the 'outside world', his description of the universe is just as correct as yours or mine.

The would-be time-traveller, in my opinion, should be discouraged from diving into the black hole. Despite optimistic speculations about hopping through a black hole to another region of space and time, or even into another universe, it seems improbable (to put it mildly) that the diver would avoid destruction. No, what he should do is to take his spaceship into orbit around a black hole, as closely as he dares, in a timeshell where his clocks—atomic and biological—run very slowly as judged by a distant onlooker. In practice, the tidal stresses would be formidably great, unless the black hole were a very large one, and he would in any case have to keep his motor running to avoid being driven in. The nearest stable orbit, where an unpowered spaceship could survive, lies at six times the radius of a non-rotating black hole; there the slowing of time is only a few months per year. But let us imagine that, against all odds, the time-traveller succeeds in skimming around and around the black hole so close to the surface that he is in a timeshell where clocks run at one thousandth of their rate at the Earth's surface. Purists please note: I am ignoring the effects of motion; in powered flight a fictional traveller can go arbitrarily slowly.

One way of judging whether or not the effect on time is illusory is to imagine signals passing between the astronaut and base. If both parties had the impression that the other's clock was running slow—as they would have, for example, were the spaceship simply travelling away from the Earth at a very high speed—then it might well be dismissed as a self-contradictory illusion. But in fact the time-traveller and his friends at base will agree on what is happening.

The signals coming from the Earth to the spaceship circling close around the black hole will appear to be running fast. In the jargon, they will show an enormous gravitational blueshift and the radio frequency of the transmissions will appear in a completely different, high-frequency band in the receiver; to understand them the astronaut will have to record them and play them back in a much-slowed form. But the information they carry will be arriving thick and fast, too. The time-traveller will receive a daily news bulletin from the Earth every 90 seconds and he will witness a US presidential election five times a week. If his friends transmit their heartbeats to the time-traveller, he will hear them as a high-pitched hum.

Going the other way, the time-traveller's signals will be registered on the Earth with an enormous gravitational redshift. His friends at base will have to tune their receivers to a low-frequency band, and be prepared to record and speed-up the messages to make them intelligible. But they will not have to work very hard at it: the time-traveller's daily reports will come in once every three years and a ten-minute greeting will take a week to record. As for the time-traveller's heartbeat, that will come to Earth about once every twenty minutes. If he sends them a television picture of himself, one blink of his eye will seem similarly protracted in time.

Yet back in his spaceship, going around and around the black hole, the time-traveller is not aware of any peculiarity about time within his own surroundings. If he feels his pulse, his heart is beating at the normal rate. In each camp, time seems to pass normally—physically, biologically and psychologically. But there also is complete consistency in the discrepancy in time between the two camps. Within a few weeks, the time-traveller will receive news that all his friends are dead. If he continues to circle the black hole for ten years and then flies back to the Earth, he will find 10,000 years have passed, into a period when his 'own' era rates only a few lines in the archaeological textbooks. But he will never be able to return to his former time. In relativity theory there is no method for any real object or individual or information to travel backwards in time.

High-speed travel, too, can keep you young. Our astronaut does not need to find a black hole in order to engage in time-travel and so outlive his colleagues on Earth. In empty space, he can plan a flight in such a way as to make the atomic clocks in his spaceship and his body run, on average, slower than the clocks at home. The plan turns out to be a very simple one: all that he has to do is to go off at a high speed in any direction, then turn around and come home again. The faster he goes, and the longer he travels, the greater his advantage in time will be.

The possibility of time-travel in empty space was among Einstein's early discoveries. It was one of the principal conclusions of Special Relativity, which he formulated in 1905, and his first strong clue that physicists would have to abandon the old notion of absolute time, ticking away everywhere at the same rate. The theory of Special Relativity is the main subject of this and the next few chapters. As its name implies, it deals with relative motion in 'special' circumstances: in empty space far removed from any strong sources of gravity. The world of Special Relativity is mathematically much simpler than the gravity-ridden world of General Relativity, but it is conceptually more confusing. Everything is footloose and there are no definite points of reference like the gravitational centres of stars and planets. I have reversed the order of presentation of Einstein's ideas, as compared with the history of their discovery: we are proceeding from General Relativity and gravity to Special Relativity and high-speed travel. Supplying links between the two theories are (1) the similarities between acceleration and gravity and (2) the effects of high-speed travel on time.

Gravity is not very strong on the Earth, or indeed in most places in the universe. As a result, the conclusions of Special Relativity apply almost exactly in all normal circumstances, and most precisely within spaceships—*Skylab,* for instance—falling freely under gravity. Although the gravitational redshift and the associated behaviour of atomic clocks may be crucial for understanding gravity, it is usually small compared with the possible effects on time of relative motion at very high speeds.

In the case of gravity slowing down clocks there is, though, a palpable cause. That a massive body such as the Earth, the Sun or a black hole should influence time and space in its environment in a peculiar way is not altogether surprising. What makes Special Relativity somewhat difficult to grasp is that, when effects of gravity are explicitly ruled out, motion alone deforms space and time. That has puzzled many eminent physicists as well as laymen.

The same is true of a central theme of Special Relativity, that the speed of light is always the same for everyone, regardless of their motions or the motions of the sources of light. This, too, is a correct though bewildering concept. A later chapter will consider it in detail. For the moment, please take it as a fact that whenever anyone travelling at a steady speed measures the speed of light in empty space he always gets the same answer, 186,000 miles a second, regardless of his own speed. With that information an effect upon time is deducible at once, taking us to the heart of Special Relativity.

If I am travelling at a high speed past you, you will judge my clock to be running more slowly than yours is. The reason becomes apparent if I am an astronaut in one of a pair of spaceships flying side by side, and you are watching us dashing past the Earth. Looking out of my window I see the other ship at rest in our little world, while the Earth rushes past in the background. But I decide to check my separation from the other ship by radar or laser beam. I send out a pulse travelling at the speed of light and time the return of the echo by my atomic clock. As far as I am concerned the pulse goes directly to my sister-ship by the shortest possible route and comes straight back to me.

That is not how it looks to you. During the short time in which the pulse is travelling from my ship, you see the other ship continuing to move forward. From your point of view the pulse has to travel at a forward-slanting angle to reach the second ship. Similarly, the returning echo has to slant forward to reach my ship, which is also still forging ahead. In other words the pulse, in your estimation, has farther to travel.

At low speeds the difference is negligible but when the ships are

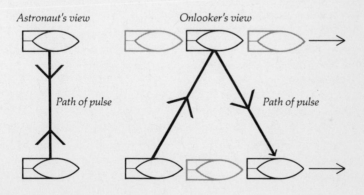

Astronaut's view   *Onlooker's view*

Path of pulse   *Path of pulse*

travelling at, say, half the speed of light you will reckon the path of the pulse to be about fifteen per cent longer than I do. But if the speed of light is the same for both of us, your estimate of the time for my pulse to go out and come back to me is about fifteen per cent greater than mine is. Yet it is one and the same radar pulse making one and the same journey and something is out of joint in our figuring of space and time.

Einstein decided that what had to 'give' in this case was the usual reckoning of time, and all subsequent investigations have justified his decision. His theory reconciles the discrepancy in distance for the travel time of the pulse by declaring that you (watching from the Earth) are entitled to suppose that my clock (in the spaceship) is running slow to just the extent needed to reconcile our two opinions about the distance travelled by the light.

Suppose that I report that the pulse echo took 100 nanoseconds (billionths of a second) to return to me—putting the ships about 100 feet apart. You will say to me: 'Aha, your spaceship's clock is running slow and it "really" took 115 nanoseconds.'

Our roles are completely interchangeable in this argument. From my spaceship I can see the Earth rushing past me, and I can watch you measuring the distance from your house to your local pub by radar. By my reckoning both you and your pub are hurrying along at half the speed of light and again your direct to-and-fro radar echo becomes a slanting, longer track in my estimation. So by my reckoning, it is *your* clock that must be running slow. The effect is reciprocal and contradictory: both clocks cannot be running more slowly than the other. This effect of high-speed motion on time is therefore in the nature of a visual illusion.

But then an amazing thing happens. If I continue past the Earth and then swing my spaceship around and fly back to you, it turns out that only the slow running of *your* clock was illusory—the slow running of *mine* was real and I have spent less time on my journey than you think. It has nothing to do with the direction of travel, because the effect of high-speed motion on clocks is quite independent of that. The distinction arises simply because I am the traveller and you are the stay-at-home.

This prediction of Special Relativity is often known as the 'twin paradox', because it is cast in the form of the story of the astronaut who leaves his twin brother on the Earth while he flies off at high speed for a long journey to the stars. On his return home he finds his twin is an old man while he, the astronaut, is still in his prime. The astronaut's clocks, atomic and biological, have registered fewer hours and years than the clocks on the Earth have done. With an appropriate flight-plan the astronaut can survive for as many Earth-years as Methuselah who, the Old Testament tells us, lived 969 years.

The twin paradox has generated more perplexity and public controversy than any other idea in relativity. Some people, hankering after more stable ideas of time than Einstein allows, find the conclusion hard to accept. In England Herbert Dingle, an astrophysicist and sometime president of the Royal Astronomical Society, put forward one counter-argument over and over again for many years. He said that, from the traveller's point of view, it is the Earth that accelerates away and then back towards him, so there should be no difference in the calendars or ages when the twins are reunited. But there *is* a difference between the two points of view.

Similarities in the situations of the time-traveller who relies upon high-speed travel and of the one who visits a black hole help to explain and confirm the twin paradox. Recall that signals from the black-hole time-traveller arrived at the Earth with a very pronounced gravitational redshift, or reduced frequency, while he received signals from the Earth with a large blueshift or increased frequency. The effects on the signals concurred and showed the time-traveller's clock to be running much more slowly than clocks ran on the Earth. For the high-speed time-traveller, too, there is a crucial period during which the Earth receives his signals redshifted while he sees the Earth's signals blueshifted.

In this case the 'reddening' and 'blueing' of the light is due to the doppler effect, which 'stretches' or 'compresses' the light waves travelling between two objects, according to whether they are moving apart (redshift) or approaching each other (blueshift). During much of the journey the effect is reciprocal, and both parties see the other 'red', or both see the other 'blue'. The crucial period comes when the traveller turns around and heads for home. At once he will see transmissions from the Earth blueshifted, and terrestrial clocks running fast. But if he is, say, ten light-years from the Earth when he turns, his stay-at-home twin cannot know that the spaceship has turned until ten years after the event, when the first signals arrive at the Earth showing that the spacecraft has turned. Throughout that ten years the stay-at-home still sees his brother's spaceship redshifted and the clock aboard it running slow. The faster the ship is travelling, the greater the difference in rates of the clocks will be and the more years the time-traveller will 'save'.

There is a more fundamental difference between the experiences of the travelling twin and the stay-at-home twin. The traveller *feels* accelerations as he sets off and turns back; the people on Earth do not. The consequences of acceleration are somewhat obscure in Special Relativity, which is why the effect might be called a paradox. Einstein himself remarked that it could be properly understood only in General Relativity. The latter theory makes it clear that accelerating in empty space is almost indistinguishable from standing in a gravitational field—and just as gravity does peculiar things to time, so does acceleration.

Whenever you travel in a loop at high speed, either straight out and back or around in a circle (the route does not matter), you might just as well be visiting a black hole. The greater your speed, the closer you are in effect going to that imaginary black hole. For example, travelling in a loop at half the speed of light is like circling a black hole at a distance of four times its radius, and you would 'gain' less than two months for every year of travel time. To match the performance of the black-hole time-traveller who slowed his clock by a factor of a thousand, you would really need to be chasing the light beams—travelling only about one mile per second slower than light itself.

So it is not simply the high-speed travel that keeps you young, compared with your stay-at-home friends, but also the exertion associated with your journey—in gathering enormous speed and in changing course to come home. In relativity theory, whether you are struggling under gravity or travelling in a spaceship, exertion can put you in a different time frame compared with the rate of clocks in 'lazy' systems. If I have laboured this account of the twin paradox a little, it is because many people besides Dingle have been puzzled by it. But in physics such issues are not settled by argument but by experiment. And when the experiment is done Einstein, as usual, turns out to be right.

To check the twin paradox on Earth it is convenient to have objects with a natural lifespan, so that you can try prolonging their lives by high-speed travel. Suitable objects come readily to hand in the subatomic world of the particle physicists. There, many particles are unstable. They have built-in obsolescence and break up after lifetimes fixed by nature. The most convenient particles for longevity experiments turn out to be 'muons'—heavy relatives of the electron. They decay into electrons after a typical lifespan ('half-life') of two millionths of a second.

How muons are made, and what they may be, are technical questions not relevant to this account. But the particles have an electric charge, which means that you can steer them with powerful magnets. Imagine, now, sending a bunch of particles off at high speed, turning them around and bringing them 'home' again. If the conclusion of the twin paradox is correct, their lifespan ought to be longer than that of similar particles at rest. The simplest procedure is to guide the muons around in a circle; the circular track also allows the survivors of one trip to go right off around again for another trip.

A merry-go-round for muons has provided the most accurate test of the twin paradox yet attempted. Emilio Picasso and an international team of physicists assembled the necessary machinery and detectors at CERN, the big European high-energy laboratory near Geneva. They built the Muon Storage Ring to check a fundamental point in the theory of the electric force, but it was admirably suited also for the relativity experiment. Constrained by magnets, particles travelling at 99.94 per cent of the speed of light circulated in an orbit 46 feet in diameter. Twenty electron detectors around the ring registered the 'deaths' of individual particles.

If its lifetime were unaffected by its high-speed journeys, a typical particle would manage fourteen or fifteen return trips around the ring before its two-microsecond life expired. In fact the typical particle in the CERN experiment survived long enough to make more than 400 orbits of the ring. Its life was extended nearly thirty-fold. Not only was the general sense of the twin paradox confirmed, but careful measurements bore out the einsteinian calculations almost exactly. In fact the experimenters were able to declare, in 1977, that the predicted factor was correct to within one part in 500. High-speed travel *does* keep you young, and Einstein's formula is commended to space-age actuaries.

14 : The Universal Correction

A passing high-speed object seems to swivel away from you.
A clock travelling at high speed seems to run slow.
An object's energy of motion increases its apparent mass.
In high-speed travel, distances will seem diminished.
Appearances change but the laws of physics survive.

When a sailing boat gathers speed under a beam wind (coming from the side) the wind shifts. It seems to come from somewhat ahead of the boat. The flag at the top of the mast, which shows the helmsman the wind direction, no longer trails exactly sideways away from the wind, but points at an angle towards the stern. The faster the boat goes the greater this angle becomes. The motion of the boat is not, of course, altering the true wind, which is set by the great areas of high and low pressure on the Earth's surface; all that changes is the direction of the apparent wind as judged by the people in the boat. Nevertheless the apparent wind is real enough: it is what drives the boat forward.

Albert Einstein was a small-boat sailor and so was the eighteenth-century Oxford astronomer James Bradley. Bradley made an important discovery in relativity, two hundred years before Einstein, as a result of reflecting on the apparent wind direction while sailing on the Thames. He realised that the apparent positions of the distant 'fixed' stars must be affected by the Earth's motion through space. Moreover, the direction of the Earth's motion changes as it circles the Sun, so the apparent positions of stars in the sky should change a little with the seasons. When he checked he found it was so—a shift of about one-ninetieth of a degree from summer to winter, for stars lying 'abeam' of the Earth's orbit. Bradley was then able to estimate the speed of light as a multiple of the Earth's orbital speed.

The astronomer has to tilt his telescope forward a little, in the direction of travel of the Earth, to catch the apparent stream of particles of light coming from a distant star. It is the same effect as the apparent 'wind' alteration of the lie of the boat's flag. (For landlubbers let me offer another analogy: tilting the telescope is like tipping your umbrella forward as you hurry through the rain, in order to keep your knees dry.) For speeds which are small compared with the speed of light, this 'aberration', as astronomers call it, is also small. But a fast powerboat can make the apparent wind come from almost dead ahead, regardless of the true wind direction. Similarly very high speeds relative to a source of light can have dramatic effects on its apparent direction and appearance.

Imagine, for example, a spaceship passing the Earth from east to west, at a speed close to the speed of light. Point your telescope towards the eastern sky and you will see the spaceship coming towards you—tail first! The aberration is now so great that the light which is to enter your telescope at the correct angle has to be launched almost straight *backwards* from the spaceship. Boats cease to be a good analogy; think instead of the spaceship at rest and the Earth hurtling past it (from nose to tail) at almost the speed of light. The only chance of hitting the Earth with a laser beam from the spaceship is to aim off almost completely in the direction of the Earth's motion—in other words, to point the laser beam tailwards.

Come back to your vantage point on the Earth: as you turn the telescope straight upwards, to try to see the spaceship at its moment of closest approach, you will still see its tail facing you. In other words, instead of facing along its line of travel past the Earth, the spaceship appears to be turned to point away from the Earth. Even at less extreme speeds, a passing spaceship will appear to be swivelled away from the Earth. You will see part of its tail when you would expect to see the ship sideways-on. Again the reason is that the light entering a telescope pointing straight outwards from the Earth has been launched somewhat backwards from the spaceship, allowing for the aberration. Many accounts of relativity say, quite incorrectly, that a passing spaceship appears unnaturally squashed or contracted along its length. It *does* appear foreshortened but only in accordance with the entirely natural perspective of an object seen from an angle.

The apparent rotation of the passing spaceship, as a result of the aberration, also gives a more direct explanation of the apparent slowing of its clock, which the previous chapter deduced by an argument about a radar echo between two ships. Because of aberration, as I have said, you see the passing spaceship by light launched somewhat backwards from it, so that it is apparently travelling away from you. As a result the light will be reddened by the doppler effect, which reduces the frequency and energy of light when the source and observer are moving apart. This occurs even when you think that the ship is at its supposed point of closest approach in passing the Earth and should be neither approaching nor receding from you. And if the apparently receding spaceship sends you time pulses from its clock, they too will arrive 'spaced out' and you will judge the clock to be running slow.

The extent of this 'transverse doppler effect' is easy to calculate, either from the 'aberration' (as discussed here) or from the apparently extended path of a radar pulse between two spaceships (pages 87–88). The normal frequency of light is greater than the frequency of similar light coming from the passing spaceship, by a factor called gamma which depends on the speed. And the apparent duration of one second on the spaceship's clock is judged by the onlooker to be one second multiplied by the gamma factor.

Speed (as percentage of speed of light)	Gamma factor
0	1·000
10	1·005
20	1·021
30	1·048
40	1·091
50	1·155
60	1·250
70	1·400
80	1·667
90	2·294
95	3·202
99	7·089
99·9	22·361
99·999	223·607
100	infinite

(The formula is gamma $= \dfrac{1}{\sqrt{1 - V^2}}$ where V is the speed of the object reckoned as a fraction of the speed of light.)

All of the subtlety of Special Relativity lurks in the transverse doppler effect, associated with the slowing of clocks in spaceships, as judged by an onlooker. To emphasise the point, let me make another table.

Direction of spaceship	Spaceship's colour-change	Apparent rate of clock	Deduced rate of clock
Towards you	'blue'	fast	slow
Away from you	'red'	slow	slow
Sideways past you	'red'	slow	slow

The distinction between the 'apparent' and 'deduced' rate of the spaceship's clock arises as follows. At first sight, the gamma factor seems to be describing a very special situation—the moment at which the Earth has the spaceship directly abeam. But in fact it operates for all high-speed motion in any direction or position. Mixed in with the ordinary doppler effect, which changes the apparent frequency of light from a spaceship travelling towards or away from an onlooker—the blueshift and redshift—there is also a small but persistent reddening effect. Pulses from an approaching, blueshifted spaceship will come thick and fast, indicating an 'apparent' speed-up of the ship's clock. But when you make allowance for this gross doppler effect, the subtler effect remains, leaving you with a 'deduced' slowing of the clock, corresponding to the gamma factor, which operates whether the source of light is moving towards, away from, or past the onlooker.

In the case of real spaceships, like *Skylab* and *Salyut,* orbiting the Earth at about five miles a second or one forty-thousandth of the speed of light, the effect is small. Ignoring the effects of gravity and General Relativity (which make the spaceship clock run fast) the prediction from the gamma factor of Special Relativity is that the spaceship clock should run slow, compared with a clock on the Earth, by about two-hundredths of a second a year.

For an object travelling at even lower speeds, gamma is almost exactly one. Therefore clocks in cars agree very precisely with clocks in buildings. For an object travelling at the speed of light gamma is infinite and time seems to stop. For intermediate speeds that are significant fractions of the speed of light, the gamma factor increases slowly at first and then rapidly, for the highest speeds.

You cannot do peculiar things to time without affecting other features of the world, and the gamma factor governs most of the curious consequences of high-speed travel. It is a universal correction. For a start, while an astronaut in a fast spaceship may appear to an onlooker on the Earth to be remaining remarkably young, he also seems to have put on a lot of weight. He thinks he is as spry as ever but, as judged by the onlooker, the astronaut's mass and that of his spaceship have increased by the gamma factor.

To understand how this comes about, remember that the mass of an object is a measure of its resistance to acceleration. Suppose that the astronaut runs his rocket motor to increase his speed. As far as he is concerned, the motor is working according to the maker's specifications and is accelerating him at, say, one *g,* increasing his speed by 32 feet per second in every second. But by the onlooker's reckoning the astronaut's second is a protracted second—gamma seconds in fact—and the spaceship is not gathering speed as rapidly as the astronaut supposes. The motor is also running more slowly, as judged by the onlooker. The effort required to increase the speed of the spacecraft by a given amount, in the onlooker's estimation, has been increased by the gamma factor.

The high-speed spaceship is seen to accelerate ever more sluggishly.

Picture the spaceship still puffing away. While it fails to gain much in speed, its mass, as judged by the onlooker, goes on increasing formidably. By the time it is going at 99·9 per cent of the speed of light, what is nominally a 100-ton spaceship by the builder's certificate has an effective mass of 2237 tons by the gamma factor. What does that extra mass represent? The energy of motion of the spaceship. The more energy that the ship acquires, as a result of its acceleration, the greater its mass becomes. As it approaches the speed of light the energy tends to become infinite (gamma = infinity) so that the spaceship's motors could work away for ever and still not generate enough energy fully to attain the speed of light. The gamma factor describes the 'light barrier', which prevents material objects travelling as fast as light.

Nor is the onlooker's estimate of the increase in mass illusory. On the contrary, if he is the one applying the force to drive an object to very high speeds, he finds that he needs to exert ever-greater effort to increase the speed by each extra mile a second. This is exactly the situation in which the operator of a particle accelerator finds himself. He pumps more and more energy into his subatomic particles but, as they approach the speed of light, they almost stop gaining in speed and pile on mass instead.

And we have come full circle, as the reader will have noticed, back to the equivalence of mass and energy, which Einstein discovered as the prime result of Special Relativity. The mass of a fast-moving object is defined by the gamma factor. The mass that remains when the object stops, and gamma reverts to one, is the rest-energy. The difference is the energy of motion. There is another echo. The slowing of the clock-rate is precisely geared to the increase in energy. Here, in high-speed flight in empty space, as in the case of gravity near planets, stars and black holes, the amassing of energy results in a slowing-down of time, as judged from the outside.

Just as time stops completely on the very edge of a black hole, so time would stop if you could travel at precisely the speed of light. You would abolish distance entirely, so that your point of departure and your destination seem to be at the same place. This becomes clear in the Irish method of measuring the speed of light. Two experimenters carefully pace out a set distance across the turf, and then Michael stands ready with a stopwatch. Patrick goes to the far end of the track with a flashlight. He raises a handkerchief, which he is to drop at the moment he switches on his flashlight. Michael will then start his stopwatch and measure the time until he sees the flashlight. With the distance travelled and the time taken, they will be able to send their results to the computer centre in Dublin and find out the speed of light. With the perfect coordination of a potential Nobel prizewinner, Michael drops his handkerchief and switches on his torch. . . .

The result is obvious. In this experiment, light seems to travel at infinite speed because Patrick sees the handkerchief fall and the flashlight glowing at the very same moment. But although they have failed to measure the speed of light correctly they have discovered something far more important, well worth a Nobel prize if Einstein had not said it first: namely that there is no way of communicating information about time more rapidly than light can travel. The particle of light carries the moment of its creation frozen into it, as it were.

If a leprechaun in a spaceship accompanies the signals at the speed of light he will think he arrives at the finishing post at the same moment as he left the start. Time 'stretches' indefinitely and distance shrinks to nothing, and the onlookers judge the mini-spaceship's clock to have stopped. Indeed someone travelling at light speed could cross the entire universe in no time at all. Space and time would be abolished. Only the sluggish movements that result from the freezing of energy into matter, and the impossibility of travelling at the speed of light, give scope to the universe and time and space for our own existence. Yet the faster you travel the more light-like your life becomes: space shrinks and time slows down.

You might in principle cover vast distances during a human lifetime. Travel to the vicinity of other stars at about one-tenth of the speed of light may become feasible for humans in the centuries ahead. At that speed any time advantage due to high-speed travel will be trivial. A journey of five light-years—a typical interstellar distance—will take about fifty years by earthly reckoning and only three months less for the travellers. In that era the idea of travelling close to the speed of light will haunt people as they make their wearisome journeys from star to star. On the basis of relativity theory (but not of any known technology) you could flit rapidly between stars, just like the heroes of space fiction.

If a high-speed astronaut has a map prepared on the Earth, which shows the positions of stars and the distances between them, he can use them as milestones to assess his speed. He may say, for instance: 'This star is four light-years from the last star, and it has taken me eight light-years to get here, so my speed across the Earth-made map is half the speed of light.' Remember that, as he gathers speed, his clock slows down, by the gamma factor. There is a magic speed (as reckoned from the Earth) at which the slowing of his clock is sufficient for him to begin to exceed the speed of light in that important practical sense. That speed is seventy-one per cent of the speed of light. At higher speeds he can go much faster than light, according to the Earth-made map: for example, when travelling at ninety-nine per cent of the speed of light (judged from Earth) he sees himself as dashing from star to star at a rate of seven light-years per year. But this is only because of the slow-running of his clock: in the earthling's estimation he is still going at rather less than one light-year per year.

And if the astronaut took the trouble to measure the distances between the stars for himself, he would find that the Earth-maps were 'wrong'. The measurements of distance depend on speed, and the whole cosmic scene is foreshortened in the direction the astronaut is travelling. If he uses radar, for instance, he will time the return of the pulses using his slow-running clock which will make the distances between the stars correspondingly less. The foreshortening exactly cancels out the effect of the slowing down of the clock and he would conclude that he was, after all, travelling at just ninety-nine per cent of the speed of light.

To make that formal correction is not to deny the astronaut's power to reach very distant places in his lifetime. Whether that possibility will ever be fulfilled, who can tell?

To see how everything changes in step with everything else during high-speed motion, look again at the electrons travelling down the two-mile-long electron accelerator at Stanford. I remarked earlier that the electrons approached very close to the speed of light and emerged about 40,000 times heavier than when they started—all of the addition being energy of motion. But the same factor diminishes the length of the accelerator. As observed when travelling at the full speed of the electrons, it shrinks from two miles to about three inches (two feet if one allows for the gradual acceleration). The electrons' 'clocks' are slowed down by the same factor, so that what we the onlookers regard as a two-mile journey would take less than a billionth of a second for the full-speed electrons. By *their clocks* and *our maps* the electrons finish up travelling about 40,000 times faster than light. But by *their clocks* and *their maps* they are still travelling a whisker less than the speed of light. By altering both distances and times to the same extent the gamma factor preserves the same speed for light.

And at last we detect the real meaning of the twin paradox, which makes it much more significant than an effective but perplexing conjuring trick. Energy is relative. For example, the high-speed astronaut is quite unaware of how energetic and weighty he has grown. In the little realm of his spaceship, he is at rest. All of the nearby stars have become, from his point of view, very swift and very massive, because of the equivalence of mass and energy. If there were no discrepancy in the clocks, the result would be somewhat catastrophic. The Earth would fall into the Sun.

If the mass of the Sun, as judged by the astronaut travelling at high speed, has increased enormously, its gravity must be correspondingly strong. Unless the speeds of the planets are in some sense adjusted, one is entitled to expect that a sufficient speed for the spaceship will make the apparent mass of the Sun great enough to drag all of the planets to their doom.

In an orderly universe you plainly cannot have the action of gravity depending on who is looking at the star—otherwise every passing sub-atomic particle might cause the collapse of entire galaxies. The slowing of clocks saves us from annihilation or, more realistically, from the downfall of physics. High-speed objects are denied the opportunity to wreak havoc wherever they go, precisely because their clocks run slow. If the astronaut sees the Sun's mass doubled, he also sees the Earth rushing twice as fast around the Sun—once every six months. That extra speed maintains the Earth in its orbit despite the apparent strengthening of the Sun's gravity.

See how neatly Einstein knits the universe together, dropping no stitches! Once he had decided to dispense with absolute time and let the clocks vary as they might, he rescued physics from contradictions of a much deeper kind. He also gained many prizes for his daring. The change in the rate of clocks, the speed of light as a barrier and the equation $E = mc^2$ are all implicit in the gamma factor. For motion at low speeds, when the gamma factor is one, the laws of motion remain almost exactly as specified by Isaac Newton. But where high-speed travel is concerned, the universe belongs to Einstein.

The laws of nature are the same for everyone—that is the sense of Einstein's Magna Carta for the universe. If it were false, the physics that lets buildings stand in winter might topple them in summer, and we might have to retune our radios from month to month. Happily we are aware that the difference in the Earth's motion between Easter and Michaelmas is 135,000 miles an hour only because the astronomers instruct us about the Earth's orbit. The fact that life goes on, undisorganised by the Earth's rotation and motion through space, is local evidence for Einstein's principle. Science would scarcely have progressed if its laws changed between Pisa and Cambridge, or between breakfast and supper-time. Beyond the Earth, astronomers can make sense of the universe and its contents only because atoms evidently behaved in *exactly* the same way billions of years ago, in a quasar now lying billions of light-years away, as they do in a laboratory on the Earth today. The universe would not create reliable conditions for life, unless there were a very large measure of orderliness of the kind implied by Einstein's principle of relativity.

One special feature of the cosmos has a grip on all the operations of every atom, star and quasar. It figures explicitly in the laws of nature concerning gravity, electricity and the sub-atomic forces. It is the ever-constant speed of light—the magic quantity c. Why is the speed of light constant? Any mild headache caused by the contemplation of that puzzle seems a small price to pay for all these benefits.

15 : The Speed of Light

In empty space, all light travels at the same speed.
The speed of light is unaffected by the speed of its source.
Einstein deduced that the observer's speed does not affect it either.
No signal or energy can travel faster than light.
For high speeds, 1 + 1 does not equal 2.

Most of the practical methods of measuring the speed of light rely upon the radar principle—bouncing light off a target at a known distance and timing its return to the starting point. Strictly speaking it should be done in a vacuum, because light slows down in glass (which is how lenses work) and even air knocks about forty miles a second off the speed of light. To check that the speed of light does not depend upon direction, you can use mirrors to deflect the light sideways, and see if any discrepancy arises. None does.

That the speed of light in empty space remains constant in most circumstances is one of the trickiest aspects of relativity to understand. It was Einstein's prime assumption and it has been fully justified by the evidence. Nevertheless it puzzled Einstein's contemporaries, and with good reason. In the case of sound waves the speed depends, just as you might expect, on how you are moving through the air that carries the sound. No such variations occur in the case of light.

Let a shaft of sunlight in through a laboratory window on the Earth and measure its speed. It is 186,000 miles a second. Now fly in a spaceship towards the Sun at half the speed of light and measure the speed of the light again (be quick—you have only a quarter of an hour before you hit the Sun). It is 186,000 miles a second. Turn around and fly away from the Sun at half the speed of light. The sunlight overtakes you at a stubborn 186,000 miles a second. Get a couple of chums and do all three experiments simultaneously; no, go further, line yourselves up so that the same sunbeam passes each of you in turn, like a thread through three beads. Regardless of the enormous contrasts in your speeds and directions, and pronounced redshifts and blueshifts, the light seems to all of you to be travelling at exactly the same speed.

Or invert the situation: make the source of light the object that is moving. Let the astronaut fly past you, flashing a laser beam at you. Catch his beam and measure its speed. While he is coming towards you at half the speed of light, you find his beam of light is moving past you at 186,000 miles a second; when he is going away, the light he flashes back at you reaches you travelling at exactly the same speed. It seems almost as if the unfortunate beam of light has to keep speeding up and slowing down, depending on who is looking at it. More than that: the same beam seems to have to travel at more than one speed at the same time. But light is just a lot of particles whizzing along minding their own business until they hit something; then they bounce off it, plough through it, or expire, bequeathing their energy to a charged particle. That is all that particles of light know: *they* never learned the equations of relativity.

Something to latch on to: particles of light are fast but careful drivers. In empty space they never overtake one another. So we can picture empty space as being crisscrossed by orderly processions of particles of light, never going faster, never going slower. Flash a torch, or explode a star, and you start another procession on its way. It takes its place in the line of traffic, among X-rays from the Sun and radio waves from a distant quasar. Whatever speed your torch or your exploding star happens to be going at, your light gets into its lane on the cosmic highway, and settles down at the same speed as the rest of the traffic. Postpone for a moment your natural question about *why* this should be so.

If all good particles of light know what speed to travel at, and keep in line, this suggests that you might judge your own speed through the universe by comparison with the speed of light. For example, you might try to adjust your motion through space until the light is going past you at the same speed in every direction, and then you will know that you are at rest in relation to all the light traffic of the universe.

Try doing that and you will find you don't have to adjust your motion; you already are at rest, and all the light traffic in the universe is going past you at the same, correct speed. 'Ah,' you say, 'Copernicus was wrong and the Earth is after all at rest in the centre of the universe!' 'Not so,' comes the astronaut's voice over the radio. 'I am at rest. The speed of light is the same in every direction for *me,* and you and the Earth are whizzing past me at half the speed of light.' A moment's thought will tell you that we have a reverse of the process which occurred when the light was launched from its various sources travelling at their various speeds. Just as it began by adjusting its speed to that of the general traffic, now at its terminus it readjusts its speed to suit whoever is looking at it. Light is all things to all men.

Einstein's universe remains democratic, as a result of all this cosmic gear-changing. It occurs so subtly that the people looking at the light are unaware that it is going on. (Secrets are best guarded when no one knows that secrets are being guarded.) In Einstein's theory there is no way of deciding on the 'true' speed of the cosmic light traffic, or who is at rest in relation to it. All observers are equal. The only circumstances in which the speed of light may seem to vary are those mentioned earlier —where accelerations or effects of gravity come into play, in General Relativity. Special Relativity deals with *steady* motion at high speeds and ignores gravity.

There are three distinct though related mysteries about the speed of light and its role in Special Relativity. The first is that light always *in fact* travels at the same speed, regardless of the speed of the source. The second: light always *seems* to travel at the same speed, regardless of the speed of an observer who can himself be travelling at a high, steady speed. Thirdly, nothing can travel faster than light.

You need not take Einstein's word for it, that light always travels at the same speed in empty space, regardless of the speed of the source. The universe offers a remarkably powerful way of checking it; all that you have to do is to watch X-ray pulsars. X-rays are better than visible light for checking the speed of light because there is much less chance of their being affected by the stray atoms in the vast spaces between us and the stars. Furthermore, X-ray pulsars move at high speed, so that any possible effect of the motion of the source on the speed of light would be greatly magnified. Nor is that the end of their virtues: they also flash very regularly, so that their speeds can be assessed with great accuracy, by the doppler effect. In 1977 Kenneth Brecher, a physicist at the Massachusetts Institute of Technology, used observations of X-ray pulsars to confirm Einstein's dictum about the speed of light more surely than ever before.

An X-ray pulsar, as I explained in an earlier chapter, is a collapsed 'neutron star' orbiting closely around an ordinary star. Gas from the latter star crashes on to the neutron star, producing X-rays. The neutron star spins, and as a result it flashes like a lighthouse. If the orbit of the X-ray pulsar is conveniently oriented in relation to the Earth, so that you see the orbit more or less edge on, the star will approach you for a while, swing around in front of its companion and then head away from you again—all at high speed.

Suppose for a moment that Einstein was wrong, and the speed of light and X-rays *does* depend on the speed of the source. The X-rays launched towards the Earth when the pulsar is approaching us might then be travelling slightly faster towards us than those launched when it is heading away. If you watched it with X-ray eyes, the effects would be very peculiar. For example, you could see the X-ray pulsar coming towards you and then *disappearing*. The reason is that the slower bunch of X-rays launched as the X-ray star is turning away would take longer to reach us. The broadcast would be interrupted, as it were, while those slower X-rays came puffing along. Over a distance of, say, 100 light-years, a discrepancy in the speeds as small as one part in a billion could cause a delay of three seconds in the arrival of the supposedly 'slow' X-rays as compared with the 'fast' X-rays. With greater discrepancies, the late arrivers could easily be overtaken by the next batch of 'fast' X-rays.

You would then be rewarded, in this conjecture, with the curious sight of a star coming and going at the same time! That may help to give you the general idea of the method, and also of how untidy the universe would be if Einstein were wrong. Brecher was looking for this and subtler effects in his X-ray pulsars—for example, apparent distortions of the orbit or delays in the eclipse as the pulsar passed behind its companion star.

Brecher's 'X-ray eyes' were provided by the *Uhuru* satellite high above the Earth. He looked at the X-ray astronomers' records for any hint of peculiarities in the appearance of three X-ray pulsars. Brecher found none and thus he confirmed that the speed of light was constant. The technique was exceptionally accurate, especially for the X-ray pulsar SMC X-1, lying 200,000 light-years away in the Small Magellanic Cloud, a nearby galaxy. He was able to say that, although the star was revolving around its companion at 200 miles per second, the distance covered by its X-rays in a second varied by less than five inches in 186,000 miles. Einstein's presumption of the constancy of the speed of light, regardless of the speed of the source of light, cannot be in error by as much as one part in a billion.

But why? For Einstein it was almost self-evident that light must always travel at the same speed in empty space. When he set down his basic assumption about the speed of light, in his earliest papers on the theory of relativity, he did so bluntly and confidently. In one case he put it in a terse footnote: 'The principle of the constancy of the velocity of light is of course contained in Maxwell's equations.'

Einstein had considered the famous equations from which the Scottish physicist James Clerk Maxwell deduced that light consisted of electromagnetic radiation of a certain speed c. Einstein concluded that the speed did not depend in any way on the motion of the source. His simplest argument was that it makes no difference to electrical phenomena whether a magnet moves past a wire or the wire past the magnet—in either case the changing effect of magnetism creates an electric current in the wire.

I should add that, thanks to Einstein, magnetism is nowadays understood to be simply a relativistic version of the electric force—an effect of charges in motion. And in electromagnetic radiation a dose of changing magnetism creates a dose of changing electricity nearby, which in turn creates another dose of changing electricity . . . and so on. This performance always progresses through empty space at the same speed —the speed of light.

By whose reckoning is the speed of light always the same? Anyone's! That the light always appears to travel at the same speed, to any observer who is himself travelling at any steady speed through the universe, is a different proposition from the foregoing one, and it was Einstein's first great inference in relativity theory. As a boy he had wondered how light would look to him if he travelled with the beam at the speed of light. He guessed that even in these circumstances the light must appear essentially normal, with the waves of electricity and magnetism advancing in the usual fashion, and not frozen in an unnatural way in space. He decided that the laws of nature, and especially Maxwell's laws of electromagnetism which fix the speed of light, ought to appear the same to any observer, regardless of his own speed.

In his celebrated paper on Special Relativity, 'On the Electrodynamics of Moving Bodies', Einstein showed how nature accomplishes this trick. The speed of light can appear to be always the same, if judgements of distances and space and intervals of time are 'private' for each observer. Thus you, on Earth, and I in a spaceship dashing past at half the speed of light can both measure the speed of the selfsame light beam and get the same answer. For example, suppose you shoot a laser beam through the rear window of my ship as I rush away from you. From my point of view, it must pass through to the nose of the ship at the usual speed of light. From your point of view, seeing that my relative speed is half the speed of light, the laser beam should take twice as long to pass through. These contradictory views are reconciled if you suppose that my ship is shorter than I think it is, and that my clock is running slow. The gamma factor discussed in the previous chapter takes care of everything.

Many experiments have been carried out to confirm that the speed of light always seems to be the same, regardless of the motions of the observer. The most famous was the one by Albert Michelson, assisted by Edward Morley, in Cleveland in the 1880s. Michelson had measured the speed of light with unprecedented accuracy. He went on to look for differences in the speed of light in different directions at right angles to each other, which might be brought about by the motion of the Earth through space. He found no difference. Historians of relativity often take the 'Michelson-Morley' experiment to have been Einstein's prime inspiration. Gerald Holton discovered among Einstein's papers a letter in which he said: 'I even do not remember if I knew of it at all when I wrote my first paper on the subject.' Yet it had caused a stir among other physicists, for whom it was the first sign that something was seriously wrong with their way of regarding the universe, and Einstein must have been aware of this new climate of opinion.

He inferred another curious effect concerning the speed of light. When the speeds of objects approach the speed of light you cannot add them together in the obvious way. Picture two galaxies rushing away from the Earth at seventy-five per cent of the speed of light, in opposite directions. Simply adding the speeds would suggest that they are travelling away from each other at 1 1/2 times the speed of light. In that case, you might think the one must be invisible from the other, because light passing between them could never catch up. But it is easy to see that they are still in contact, in principle. For example, one of them could send a message to the other, if need be by way of the Earth. The speeds of the galaxies relative to the Earth do not affect the speed of a signal.

Sitting on the Earth we could receive a signal from galaxy A that reads: 'Warmest greetings on Einstein's birthday. Please pass on to galaxy B.' So then we send off a message that reads: 'Galaxy A sends you greetings on Einstein's birthday.' We know that it can eventually get to its destination because we can *see* galaxy B. But even if we and the Earth were not here (or were asleep when the message came) you can still imagine galaxy A's message whizzing past the Earth's position in space without any intervention on our part, and eventually arriving at galaxy B. So adding the speeds gives the wrong answer: the speed at which A and B are moving apart must to them seem *less* than the speed of light, otherwise no such message could pass.

What is the explanation here? We have to figure out what the speed of galaxy B seems to be from the point of view of galaxy A. If that came out at something greater than the speed of light, then the two galaxies would indeed be mutually *incommunicado*. To find the answer, the relativist divides the simple sum of the speeds by a certain factor (not the gamma factor, but something similar) which takes account of the slowing down of time, as judged by us, in the two galaxies.

Physically the situation resembles that of our astronaut who thought he was flitting between the stars faster than light. Distances in the universe are foreshortened for beings in the high-speed galaxy A, compared with our impression of the distances; accordingly their judgement of the speed of the distant galaxy B is reduced. The correcting factor in the case of the two galaxies moving at seventy-five per cent of the speed of light in opposite directions is about 0.64. Multiply the crude addition (150 per cent) by that factor and you find the mutual speed of galaxies A and B to be ninety-six per cent of the speed of light.

It is interesting to note that, if the galaxies seemed to us to be travelling at 100 per cent of the speed of light in opposite directions, they would seem to each other to be travelling apart, not at twice the speed of light but at exactly the normal speed of light. That leads naturally to our third 'mystery' about the speed of light.

Why does nothing travel faster than the speed of light? Physicists can easily imagine 'things' (waves, for instance) going faster than light, but they are generally persuaded that no meaningful signals or supplies of energy can go faster than light. It is useful to start with some exceptions that help to 'prove the rule'. Astronauts and galaxies can, as we have seen, cover distances judged by us, in times judged by them, at a rate giving the impression that they are travelling faster than light. The foreshortening effect on distances, though, means that they are travelling more slowly than light, when they measure the distances for themselves.

At a more mundane level, objects can travel faster than light, when light travels slowly through water or other materials. An eerie blue glow in the water of a 'swimming-pool' nuclear reactor is due to shock waves of light, set up by electrically charged subatomic particles travelling through the water faster than light—faster than light travels through the water, that is. The effect is very similar to the sonic boom (a shock wave of sound) set up by a supersonic aircraft, which in turn is closely akin to the bow wave (a shock wave of water) set up by a ship. In the case of light, the rays so produced are called Cherenkov radiation, after the Russian scientist Pavel Cherenkov who discovered the effect in 1937. Routinely, high-energy physics laboratories employ detectors based on the Cherenkov effect to gauge the speeds of particles. But, while the particles can go faster than light does *in water,* they are still travelling more slowly than light travels in empty space.

The reason why ordinarily nothing can travel faster than light in empty space can be given from two different points of view. The first is the physical and mathematical. As discussed earlier in this chapter, anything attempting to 'crash the light barrier' or even just accelerate up to the speed of light finishes up by acquiring more and more energy but very little additional speed; it becomes progressively harder to accelerate. Going a little deeper, we can say that objects up to and including the speed of light have ordinary energy, which is associated with ordinary time. At higher speeds they would have 'negative energy', and it is not obvious what that can mean except a gross affront to our ideas about time.

That introduces another, more philosophical viewpoint. If objects could go faster than light they would evidently be going 'backwards in time'. For an impression of what that means, imagine watching an object coming towards you faster than light. You would see it apparently going away from you, back on its tracks. It would appear near to you, at the end of its journey, before the light arrived from farther off, telling of its coming.

This effect is well known in relation to the much slower speed of sound, with supersonic bullets and aircraft. But in the case of light, the motion of which is closely coupled with the fundamental workings of time in the universe, the consequences are serious. For example, if you were shot, you could in theory send a signal faster than light and backwards in time, which would stop the person pulling the trigger that fired the shot which had already hit you. In other words, all our experiences about cause and effect, and the natural flow of time, could be brought to nought by signals—or other forms of energy or matter—going faster than light. Einstein's rule again keeps the universe simple and tidy.

A possible loophole in the theory of relativity nevertheless led some physicists to consider whether particles of a special kind might not exist on the other side of the light barrier: 'tachyons', as Gerald Feinberg of Columbia University named them. The idea was that, from the instant of their creation, they went faster than light and always did so. They could not cross the light barrier from the other side, so they could not slow down nor enter into normal courtesies of sub-atomic existence. In mathematical terms, their energy would be 'negative' and their rest-energy or mass 'imaginary'. If tachyons carried an electric charge they ought, though, to produce Cherenkov's shock waves of light, and might be detectable by that means. Experimenters have looked for tachyons that way, and found none.

Alternatively, even if tachyons had no electric charge it might be possible to infer the creation of tachyons during reactions between ordinary sub-atomic particles, by discrepancies in the total energy. The expectation becomes a little weird: namely that the ordinary particles coming out of a sub-atomic reaction would have more energy than the particles that went into it—because the creation of tachyons with negative energy would donate positive energy to the other particles. Tachyon hunters laboriously scanned thousands of records of sub-atomic interactions. Again they found nothing.

I mention the tachyons, not to inspire any hope for their existence or detection, but to illustrate once again the importance that physicists attach to checking every possible aspect of Einstein's theory. The loophole still exists, in principle if not in practice, and particle physicists are in the habit of thinking that anything that is not expressly forbidden by nature is compulsory. It is not enough to say—as one well might—that the universe would be a disorderly place if tachyons could keep injecting mysterious supplies of energy, and travelling backwards in time. Some theorists will not be satisfied until Einstein's reasoning is extended to show precisely why tachyons cannot after all exist.

Meanwhile, Einstein's presumptions about the speed of light are verified indirectly in all of his predictions, from $E = mc^2$ to the subtle effects of einsteinian gravity. For him it was a matter of intuition; for modern physics and astronomy it is bedrock.

16 : Where Time Flies

All observers can agree on the separation of events in spacetime.
The distance travelled by light represents time.
A 'world line' in spacetime shows the history of an object.
Acceleration and gravity bend the world lines.
A postscript: space is not empty, but not syrupy either.

Albert Einstein did not cut adrift all the pieces of the universe to wander vaguely through ill-defined space and time. While he abolished the framework of absolute space and absolute time he replaced them by absolute spacetime. Although a more malleable kind of thing than its predecessors it is, if anything, more reliably absolute. Newtonian space and time were like a chessboard on which the game of matter and energy was played. Einsteinian spacetime takes an active part in the game and may even be regarded as being created by energy. At any rate, it shares fully in the history and fate of the contents of the universe. Among the paraphernalia of relativity theory, the speed of light is absolute, the pattern of curved space near a massive object is absolute, the rest-energy of an object is absolute. And all the laws of physics are absolute, not in the sense of being unalterable by the progress of research but in the sense I have already noted—they are consistent throughout the universe.

The difficulty about spacetime is visualising the mixture of space and time. To say a little about how relativity theorists think about spacetime may be illuminating. Some of their terms, such as 'event', 'world-line' and 'light-cone', have a certain vividness in their own right. They may help the reader to a sharper sense of how space and time interact in Einstein's universe.

An 'event' means essentially what it does in everyday parlance: something happening at a particular place and a particular moment. Four figures will pin down an event—three to specify the position in space and one for the time. If, for example, passenger Brown takes a sip of champagne at latitude 46°55′ north (first figure—*space*), longitude 7°28′ west (second figure—*space*), with his aircraft at 30,000 feet above sea-level (third figure—*space*), at 20 seconds after 6·14 p.m. local time on 14 March 1979 (fourth figure—*time*) he can define an event in spacetime, while he toasts Einstein's birthday looking down on the city of Bern.

Such a scheme can in principle specify any event in the universe, given suitable maps and calendars. So far, nothing very mysterious. We can go on to say that passenger Brown's plane landed 430 miles away from Bern, 58 minutes and 43 seconds after he sipped his champagne. That is the interval between two events. But here the relativist looks dubious. 'By whose reckoning?' he wonders. The Bern control tower cannot know that the plane has arrived safely at Rome until a message has been received, which will take at least 0·0023 seconds at the speed of light.

While admitting that this may seem a pernickity point in terrestrial circumstances, the relativist sees grave difficulties looming up beyond the Earth and tells us that he would sooner reckon the 'interval' between the events in a way that combines space and time. This he does by converting the time figures into space figures, as the distances that light would travel in the given times; then he treats time, so reckoned, almost as if it were an extra dimension of space. On this basis the trip from Bern to Rome was more like 660 million miles, which is the distance light travels in an hour. For the relativist the great advantage of defining the 'interval' in this apparently idiosyncratic way is that it is the *least* idiosyncratic way of doing it. Even an observer in a distant galaxy will agree that that was the 'interval' between the two events in passenger Brown's life.

In this unfamiliar way of reckoning intervals we symbolise the passage of time by imagining passenger Brown hurtling along at the speed of light. He is shooting upwards in one dimension as it were, at the same time as his aircraft flies 'sideways' in other dimensions. I must emphasise that there is no realistic sense in which he is travelling at the speed of light, and add a word of caution against being carried away by the idea of time as the 'fourth dimension'. It sometimes is regarded as an occult extension of the familiar three-dimensional world. It is nothing of the kind: it is simply time, treated as an extra dimension for mathematical convenience. Space and time remain distinctive kinds of things and are treated differently in the equations of relativity. Nevertheless time and space do interact, and the real consequences are quite fascinating and puzzling enough without inventing supernatural mysteries about them.

The symbolic way of taking account of time is very useful for relativists, and it gives rise to the notion of the world-line, which is the track of an object through a four-dimensional world of space and time. We can, for example, depict Albert Einstein's own world-line, if we flatten the surface of the Earth and use the vertical to represent time. We ignore the motions of the Earth. As the various cities where Einstein lived did not move about, their world-lines are like a set of parallel railway tracks through spacetime, and Einstein is like a train switching from one line to the other. He is born in Ulm, in Germany, and hurtles upwards through one light-year's worth of time (i.e. one year) before his family switches to the world-line of Munich (for 13 years) then to Pavia and so on through his career: Zurich, Bern, Berlin . . . There are little wobbles as he goes sailing at Caputh near Berlin, or travels to meetings. Eventually he emigrates to Princeton and, by the time of his death there at the age of 76 years and 34 days, he has notionally travelled through time for a distance in light-years somewhat greater than the distance to the bright star Aldebaran.

That is the idea of the world-line: using it for physics rather than biography, the relativist pictures the universe as an untidy haystack of world-lines, corresponding to all the particles of matter, big or small, that are whizzing about at different speeds and in different directions. If two particles collide and rebound their world-lines first converge and then bend away from each other.

The world-lines of particles of light are something special—as you might expect, seeing that the speed of light is used to set up the framework for the world-lines. If we continue compressing three-dimensional space to two dimensions and reserve the third dimension for time, the light usually travels at an angle of 45° to the flattened space. In this scheme, all of the light arriving at a given point, from any direction in space, makes a cone converging on the point; similarly all the light leaving the point in question is diverging from it, again as a cone. At that point, 'here and now', the knowable past is defined by the light-cone converging on the point from 'behind'.

No information can reach you except from the realm of spacetime lying within that light-cone. You cannot know whether the star Betelgeuse is blowing up 'today' because its 'today' lies outside our light-cone, 600 light-years away. Six centuries from now when your descendants have moved sufficiently far 'forward' in time to bring that possible event within the light-cone in their wake, they will know whether Betelgeuse did blow up 600 years before. (In Betelgeuse's present light-cone, the Earth appears as it was 600 years ago, when the Mongols were on the rampage.)

Similarly another light-cone opening up 'forward' in time defines your possible future. You cannot visit Betelgeuse tomorrow because you cannot go faster than light. You could flash a signal to it that would arrive in 600 years, because that is when the world-line of Betelgeuse will intersect your future light-cone. The light-cones may seem reminiscent of the 'light-bubbles' described in an earlier chapter. They are the same thing, but now represented as cones on the spacetime diagram. The widening of the light-cone is simply the expansion of the light-bubble as time passes.

If you set off yourself to visit the star at one-tenth of the speed of light you would follow a world-line lying well inside the light-cone, making an angle of 6° with the central (time) axis, and arrive at Betelgeuse in 6000 years. If the time-axis of the cone represents the world-line of the Earth, you, like all other high-speed objects in its vicinity, are now following a slanting route through spacetime. In relation to the Earth your path is therefore somewhat more like that of light, going off at 45°. At the speed of light, time would stand still. The faster your speed, the closer your angle will approach 45° and the slower your clock will be running, by comparison with the clock on Earth.

The world-line of the travelling twin in the twin paradox first diverges from the world-line of his twin on the Earth and then curves back to converge with it again. The parting and reunion are the two points where the two world-lines meet. They have followed very different routes in spacetime and so for the relativist it comes as no surprise that their clocks do not agree. The non-relativist may scratch his head

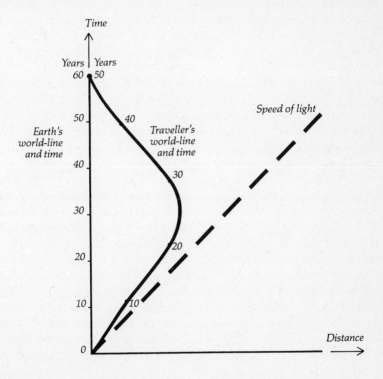

over the fact that the travelling twin's route looks longer on paper, although it is shorter in time. This is an oddity of the spacetime diagram. The distinguishing feature of the travelling twin's track is that it is bent: the bending, a change in direction and for speed, corresponds to an acceleration.

Acceleration feels like gravity. Another way of imitating gravity is by rotation, which is a special form of acceleration. Even at a constant speed, the rim of a rotating object is continuously accelerating towards the hub, while going sideways quickly enough to remain at a fixed distance from it. The ability of potential astronauts to withstand the g-forces of their profession is tested by putting them in a centrifuge— a fearsome merry-go-round in which they whirl at a high rate and feel the g-forces acting outwards. This 'centrifugal force' is indistinguishable from strong gravity, except in so far as the passenger may feel giddy. And scientists who propose building giant settlements in orbit around the Earth promise to provide Earth-like conditions by rotating each space settlement at just the right rate to provide one g at the perimeter; instead of being weightless, objects will seem to behave 'normally'. In the spacetime diagram, the world-line of a person in a centrifuge travels in a helix around the world-line of the hub of the centrifuge. Because the direction of his world-line is continually changing, he feels the continuous acceleration.

In the case of gravity itself, the paths of light are altered when they pass near to massive bodies. Gravity tilts light-cones a little inwards towards a massive object like the Earth or the Sun. (In our earlier terms, the light-bubble is displaced off-centre.) And time is slowed by the deflection of the light-cone. The all-important angle between one's world-line and the light is reduced, not by high-speed travel but by the effect of gravity on light.

At the edge of the black hole, the cone is tilted through the full 45°, bringing its outermost edge in line with the world-line of the edge of the black hole. If you are at the edge of the black hole, your future defined by your light-cone now lies entirely within the black hole. Once inside the black hole, you will find that the cone goes on tilting more steeply yet. Your future shifts inexorably nearer to the centre of the black hole. Relativity predicts that the contents of a black hole should crowd together in one geometrical point at the centre—the singularity. All the world-lines converge at that point and stop.

Within the scheme of relativity acceleration is certainly detectable. The gravity sensors in your ears respond to acceleration by motors, by centrifuges or by gravity. What you are not supposed to feel or detect is any effect whatever of steady motion in a straight line in empty space. But according to the quantum theory, space is never empty. In a postscript to Special Relativity, we should take account of the ghostly particles of light and matter that 'seethe' even in the purest vacuum. Given their existence, it is not at once self-evident that you will be unable to detect your steady motion through space.

At first sight you might well think that this underworld in space will look different, depending on the speed at which you are travelling. After all, the ghostly particles will have added energy, from the point of view of an astronaut who is going at a steady speed. Can he not therefore tell himself that he is at rest while everything is rushing past him? 'Aha,' he says, 'all the atoms in my spaceship will be retuned a little, depending on how fast I am travelling. I can make a cosmic speedometer to tell me my time speed through space.' This would easily destroy the theory, if he were right.

You would expect the ghostly light to be more energetic ahead than behind—as in the case of light from the Sun when the astronaut whizzes past it. As a result there ought to be a pressure on the front of the spaceship greater than on the back, tending to slow it down. Objects should not then persist in their motions but ought to grind to a halt in the syrup of space.

To try to avoid this catastrophe, for Newton's theory as well as Einstein's, you can ask a question. Is there a special population of ghostly light particles that will look exactly the same to you regardless of what speed you are travelling at? The answer is yes: one and only one: for each doubling of the light frequency there must be eight times as many particles of light. It turns out that this population of particles is what the quantum theory predicts—exactly as required to safeguard relativity. You might shrug and say: 'But we know that the Earth has marched unimpeded through space for billions of years, so it must be all right.' The physicist is more awestruck by the ability of nature to arrange such things just so, and he looks for deeper explanations of why space is so conveniently non-viscous.

Having sung the praises of Einstein's theory, and told how it was saved from drowning in the syrup of space, I must now prepare the philosophically minded reader for a nasty shock. Special Relativity is not, after all, strictly correct! The reasoning we have followed is, I believe, impeccable in so far as it describes what goes on in respect of light, time and so on in spaceships. What is false is nothing less than one of Einstein's fundamental assumptions: that it is impossible for an astronaut moving at a steady speed to tell whether he is moving or the outside world is moving. In fact it turns out that he can, and the cosmic democracy of Einstein's theory is compromised. To adapt George Orwell's slogan from *Animal Farm:* all observers are equal but some are more equal than others.

Fortunately that does not seem to affect any of the conclusions of Special Relativity already presented. The recent discoveries do, though, give us back something not unlike the absolute frame of space that Einstein thought he had abolished. In this respect, at least, Einstein's universe recovers a little of the flavour of Newton's. How that comes about will emerge in the next chapter, when we find out by what trick our busy astronaut is able to put together that cosmic speedometer, using the Big Bang itself.

17 : The Simple Universe

The galaxies are flying apart as if from an explosion.
Radio energy pervading the universe is a relic of the Big Bang.
In General Relativity the simplest universe explodes and then collapses.
The size and duration of a Simple Universe are fixed by its energy.
Einstein could have predicted the Big Bang, but didn't.

In terrestrial affairs we think of big as being complicated: a city more intricate than a village, an ocean more complicated than a puddle. For the universe the reverse seems to be the case: bigger is simpler. Living things, far too small to register on any cosmic scale, are exceptionally complicated and subtle. Planets, too, can be extremely complicated, as any textbook of geology will convince you. Stars are fairly complicated, especially when you look at them very closely (as we can look at the Sun) or when you trace their entire life cycles.

Galaxies have some puzzling features: for example the elegant spiral arms (as in our own Milky Way) or the violent nucleus (like that of M87). But on the whole they are scarcely more complicated than the stars which compose them. Beyond the galaxies, in the hierarchy of the cosmos, there are clusters of galaxies. These are very simple, as far as we can tell, just a large or small number of galaxies loosely kept in company by the gravity of their largest members, the flagships of the clusters. But simplest of all—astoundingly simple—is the universe at large. In its broad features, it is far less complicated than the Earth, one of its most trivial constituents.

Present knowledge about the universe can be accurately outlined in three sentences. (1) It consists of billions of galaxies flying apart as if from an explosion. (2) Apart from the luminous galaxies the universe is filled with radiation which is interpreted as a relic of the explosion with which the universe began. (3) The galaxies and the radiation look very much the same in all directions and the universe is not lumpy, lopsided or rotating.

This simplicity is almost certainly not an illusion created by ignorance. On the contrary, the more thoroughly the astronomers investigate the universe, and the more closely they check for possible complexity or disorderliness, the more clearly the simplicity shines through. There are, of course, unanswered questions about the universe, some of them deep and perplexing, others tantalising because the answers seem almost within the astronomers' grasp. But the answers can be sought within a framework that seems very clear. Indeed, one of the deepest questions of all may be: 'Why is the universe tidy?'

Albert Einstein's relativity has become the means of comprehending the universe at large. Not only did he discover the energy of creation ($E = mc^2$) and the modern laws governing the behaviour of massive bodies and high-speed objects in it, but his equations incorporate the origin and fate of the entire universe. Before examining the possibilities that they describe, we should first note how the main features of the universe revealed themselves to astronomers.

Amid all of the twentieth-century disclosures about the constituents of the universe—galaxies, radio galaxies, quasars, pulsars, X-ray stars and so on—two discoveries dominate modern cosmology, the science of the overall nature of the universe. They are not details but comprehensive features. One discovery dawned gradually, the other came like the proverbial bolt from the blue.

In the 1920s and 1930s, Edwin Hubble sat at the 100-inch telescope on Mount Wilson near Los Angeles night after night, year after year. For his painstaking research into motions of the galaxies, he invented ways of estimating the distances of galaxies and used Doppler's effect to judge their speeds. The light of a distant galaxy was always redshifted —reduced in frequency—indicating that the galaxy was moving away from our own Galaxy, the Milky Way, at a high speed. This was remarkable enough; it implied a general expansion of the universe. Even more remarkable was the rule he discovered about the relationship of speed and distance. In 'Hubble's Law' the speed of the galaxies was in simple proportion to their distance: double the distance and the galaxies were going away twice as fast. By 1929, Hubble had established his law out to a distance of six million light-years.

When this law emerged the implications were startling. Hubble's law was exactly what you would expect to find if all the galaxies had started off from one place and then moved apart, with the fastest galaxies reaching farthest afield. In other words, if you traced back the motions of the galaxies, they were all massed together at a certain time in the past when the universe must have been in a state very different from the present state. The theory of the Big Bang was the simplest and most natural interpretation of this discovery.

There was a big snag. Hubble estimated the maximum age of the universe since the start of the expansion—the 'Hubble time'—at two billion years. Assuming that gravity had slowed down the galaxies since the expansion began, the universe had to be younger than that. But there was already solid evidence that the Earth was several billion years old and many stars seemed much older. How could the Earth be older than the universe?

For this good reason, the Big Bang theory was treated with considerable reserve until the 1950s, when Hubble's former pupils began to discover flaws in his scale of distances. Since then the Hubble time has been revised upwards; nowadays it is generally taken to be about 15–20 billion years, or ten times Hubble's original estimate. Allowing for a slowdown, that puts the start of the expansion at 10–15 billion years ago, which accommodates the ages of the oldest known stars (about ten billion years) and of the Earth (now put at 4·55 billion years).

Direct evidence of the Big Bang came in 1965, ten years after Einstein's death. It took the form of 'stray' radio energy in the universe, not associated with individual radio galaxies or quasars. Arno Penzias and Robert Wilson of the Bell Telephone Laboratories discovered it during tests connected with the earliest communications satellites. As far as a telephone company is concerned even radio energy from the universe counts as 'static' or interference on communications circuits. It was while they were experimenting with a large horn and a very sensitive receiver designed to pick up radio energy of extremely high frequency (microwaves) that Penzias and Wilson detected weak but unceasing 'static' filling the whole sky. They did not know what to make of it until they heard, from nearby Princeton University, that the Big Bang could be the explanation.

To cut a long story short: empty space is filled with energy corresponding to a temperature of about three degrees ('3K') above absolute cold. The 3K radio energy is the present form of radiation from a great flash that occurred when the universe was very young—possibly visible light, originally as white as the Sun's but now immensely redshifted by the subsequent expansion of the universe. You can think of the waves of light being 'stretched' and their frequency reduced, in just the same proportion as the universe has grown since the flash. Because the radiation has hurtled freely through the universe for most of the time since then, astronomers have an uninterrupted view of a late stage in the primordial explosion. The quantity of radiation is immense. For every atom of hydrogen in the universe there are about a hundred million particles of 3K radio energy, and their total mass-energy is about one-thousandth of the mass of the galaxies.

Microwave astronomers looked very carefully but for more than ten years they could detect not the slightest variation in the intensity of the 3K radio energy; in whatever direction they pointed their detectors, it varied by less than a few parts in a thousand. Dennis Sciama calls this 'by far the most accurate measurement ever made in cosmology' and it provided very strong evidence for remarkable uniformity in the universe at large. It virtually ruled out any possibility that the universe as a whole is rotating. It also implied that our Milky Way Galaxy was taking an orderly part in the general expansion. If it were travelling in a maverick fashion at any appreciable speed, the energy would vary noticeably in different directions in the sky. This brings me to what I hinted at in the previous chapter: the 3K radio energy pervading space provides a means of measuring a steady speed in relation to the universe at large, in apparent defiance of the spirit of Special Relativity.

Our astronaut can make a cosmic speedometer consisting of one or more horns for detecting the 3K radio energy and measuring its intensity with great accuracy. If he now travels off in any direction at high speed, the microwave energy will appear, by the doppler effect, more intense in the direction in which he is travelling, and weaker in his wake. He can thus establish his speed. If, for example, he is going at one hundredth of the speed of light, the energy of the microwaves ahead of him will seem intensified by one per cent. If he increases his speed to half the speed of light, the sky in front of him will glow visibly red and he will feel its heat.

Experimenters from the Lawrence Berkeley Laboratory in California made exactly such a cosmic speedometer for the Earth. In 1976-7 they flew it on a U2 aircraft, high in the Earth's atmosphere. Richard Muller and his colleagues were aiming to measure the 3K radio energy in different directions in space with greater precision than ever before. (The motion of the aircraft carrying the microwave detector was, of course, utterly insignificant compared with any cosmic motion of the Earth.) They were confident that they would detect *some* motion: after all, the Sun orbits in the Milky Way, at about one thousandth of the speed of light. The Earth cannot be completely at rest amid the remnants of the Big Bang.

They found that the intensity of the 3K radio energy was strongest in the direction of the constellation Leo—the sky was about a thousandth of a degree warmer there than elsewhere. But the effects of the motion of the Earth around the Sun and of the Sun in the Milky Way have to be taken into account. It then turns out that the Milky Way— according to these results—is cruising through the universe at 1/500 of the speed of light (almost 400 miles a second) in the direction indicated by the constellation Hydra. Balloon observations of the 3K radio energy by Brian Corey and David Wilkinson of Princeton suggested a slightly lesser speed. In either case the speed is faster than cosmologists expected; it implies that the Milky Way and all its neighbours (such as the famous nearby galaxy in Andromeda) may be influenced by the gravity of a 'supercluster' of more distant galaxies.

What exactly do I mean by 'cruising through the universe'? There is a cosmic frame of reference defined by the 3K radio energy which shows us, albeit in expanded and redshifted form, the universe as it was ten or fifteen billion years ago. If you think of any group of fragments of matter, they will have moved apart to much greater distances, and the continued expansion ensures that the remote fragments appear to be travelling away very rapidly. But at any locality there is a standard state of 'rest' relative to the universe at large. In a spaceship, you could adjust your motion until you see the 3K radio energy to be *exactly* the same in all directions; then you are at rest. 'Cruising', by contrast, means travelling relative to that ideal state of rest. Cosmic theorists are looking for more detailed measurements of the 3K radio energy, that may reveal unevenness across relatively small regions of the sky. That would be evidence of 'lumpiness' in the Big Bang. If it were perfectly smooth, the theorists could not explain how the later agglomerations of matter occurred, which created the galaxies and clusters of galaxies.

There is another sense in which cosmology may offer something not unlike the pre-einsteinian framework of space. It possibly clears up the Newton Mystery. Newton had good relativistic instincts yet found himself forced to believe in absolute space. The reason was that, when he spun a bucket of water as it hung from a string in his laboratory, the water heaped up at the rim. Somehow the water 'knew' whether it was rotating or not—as if there were some absolute frame of non-rotating space with which it could compare its motion. Einstein rejected absolute space, but Newton's experiment still works. In one interpretation of General Relativity, the effect on the water in the bucket results from the action of all of the rest of the universe. It is as if we live inside a hollow, massive sphere composed of the stars and the galaxies.

You are quite entitled to suppose that the spinning bucket is at rest and the whole of the universe is rotating around it. But, if so, you have to take account of the fact that, in Einstein's theory of gravity, a rotating mass has a different effect on space and time from one which is not rotating. Recall the spinning black hole that dragged space around with it, in a fearsome carousel. The great shell of the supposedly rotating universe would have a related effect and would cause the water in the supposedly stationary bucket to 'fall' towards the rim of the bucket. I should add that theorists are divided about the merits and importance of this explanation.

On the other hand the remarkable uniformity of the universe as a whole permits most theorists to agree upon the course of events during the Big Bang and to describe them with remarkable assurance. They tell of a frenzied mass of radiation, matter and anti-matter, expanding and cooling. In the first few minutes most of the matter was annihilated by the anti-matter, but a relatively small residue remained. Eventually the universe cooled sufficiently to allow the particles to come together as atoms of hydrogen and helium—the raw material of the universe. A great flash of light occurred, as the atoms formed, and that was the possible origin of the 3K radio energy.

By a process which is not entirely clear, the atoms massed into galaxies and stars. The universe went on growing, with the galaxies moving farther and farther apart. In short, the universe began as an extremely dense accumulation of energy, and then grew. The discoveries that revealed this picture were extraordinary. Just as extraordinary was the fact that, when theorists came to consider how to describe such a universe, they did not have to look very far. The whole story was already implied in Einstein's theory of gravity. In relativistic cosmology the energy of the universe creates space and time.

When physicists and astronomers play at being God, they try to imagine an overall 'design' for the universe which encompasses the origin and fate of all the atoms, stars and galaxies within it, yet avoids trivia like the origin of the Sun and the Earth. Nobody ever played this game more skillfully than Albert Einstein or botched it so badly. He lost his nerve at the critical moment in 1917 when the oracle of his mathematics confided to him a cosmic story that he found altogether too sensational to believe.

The man who had dared to monkey with time and to write $E = mc^2$ shrank from laying bare the beginning and possible end of time. He might merely have written $E = 1.18 \times D$. In that very simple equation E is the total energy of the universe, and D is the maximum diameter of the universe. Ordinarily, energy and distance are measured in different units, but for this purpose they can be converted, one into the other. (For example, the distance associated with one unit of energy E turns out to be a quarter of the diameter of a black hole of mass or rest-energy E. The number 1.18 comes in for incidental geometrical reasons: it is three-eighths of 'pi'.) Leaving the details aside, we can say that Einstein's pristine theory of gravity shouted an extraordinary possibility. The maximum diameter of the universe may depend on its total energy and nothing else.

Nor was that all. The 'maximum diameter' represents only a moment in the cosmic story. It is like the maximum diameter of the lung of a deep-breathing man: something that can be attained for a moment but not held. The theory of gravity makes a restless universe virtually unavoidable: it has to be either expanding or contracting because, if all its contents were scattered but static, their mutual gravity would at once start hauling them together.

Still in amazingly simple mathematics, the theory fills out the story of what I shall call the Simple Universe. It starts very small and spontaneously grows in diameter extremely rapidly. Thereafter the rate of expansion of the universe diminishes steadily as it approaches its maximum diameter. Then it stops growing and begins to collapse. It shrinks faster and faster, until once again it is extremely small, and all its contents are destroyed in a Big Crunch. Einstein's equations provided a modern version of *Genesis* and *Revelations.*

The most uncanny feature of the story is that not only the maximum diameter of the universe but the entire time-scale of events is fixed only by the amount of energy in it. Imagine a Simple Universe with energy equal to the rest-energy of the Sun. The equations say that this petty cosmos will grow spontaneously to a maximum diameter of rather less than a mile (about 1370 yards) and then collapse in an equal time. Every Sun's-worth of energy that we add to the recipe for the universe will add about four light-microseconds to its maximum diameter and 6.6 microseconds to its total lifetime. To make its lifespan equal to a human being's 'three-score years and ten' requires energy equivalent to 300 million million stars; in that sense each of us owes the time of his life to the existence of about a thousand galaxies.

Suppose, as some cosmologists do nowadays, that the total energy of our universe is about 3×10^{23} (3 followed by 23 zeros) times the rest-energy of the Sun. Then, by Einstein's simplest formulae, it is due to expand to a maximum diameter of about 40,000 million light-years. Its total lifespan from its explosive genesis to its compressive doomsday is, by this reckoning, about 63,000 million years. From other information we can estimate that we are living at a time one-sixth of the way through the universe's life cycle, when it is still expanding rapidly. These estimates were not available to Einstein in 1917, but the general story was there, in his equations.

If only Einstein's nerve had held, here would have been his master stroke. He would have predicted the recession of the galaxies that Edwin Hubble announced ten years later and, long before anyone else, he would have promulgated the Big Bang as the origin of the universe. But in 1917 no one imagined that the universe was anything like that. They thought it was pretty static. Indeed Einstein feared for the safety of his whole theory when it depicted the cosmos in so different and apparently absurd a way. Therefore he deliberately 'bent' his theory, in order to force it to describe a less melodramatic sort of universe.

Einstein was full of remorse later, and called that tinkering the biggest blunder of his life. As we shall see, it is not impossible that some tinkering will be needed, when astronomers are better able to judge the long-term future of the universe. But, before considering possible variations, let us explore the most rudimentary of einsteinian universes a little further. The Simple Universe is a fascinating concept in its own right and it provides a basis for discussing the alternatives. What is more, it may prove to be precisely the kind of universe that we inhabit.

I have described the universe swelling up like a lung and then contracting again. Sceptical astronomers will sometimes tell you that we cannot see the *universe* expanding: we can only see galaxies and quasars moving apart. In the sceptics' view, there may well be empty space already existing 'out there' and the galaxies are moving into it. (The comedian Peter Cook captured a semantic difficulty for such scepticism when he remarked: 'I am specialising in the universe and all that surrounds it.') For the Simple Universe, it is not like that. There is no space outside the universe, and no time either, unless they belong to completely separate universes. Time began with the Big Bang which created it and will end with the Big Crunch. Space too is created and defined by the contents of the universe.

The spaces between galaxies grow, but atoms and galaxies do not. If they did, we should not know of any growth because we and all our instruments would be growing too; that would be possible only if the forces of nature grew weaker with time, which General Relativity forbids. As for our vantage point in space and time, we are in the midst of the universe, so we do not see the Big Bang as if we were looking at it from 'somewhere else'. It has happened all around us, leading to the curious inversion that the universe in its most compact state seems to be the biggest thing we can see. Our position is like that of a beetle in a currant loaf that is expanding in the oven: the currants, representing other galaxies, move farther away from us on all sides. We, or any other beetles, have the illusion of being at the dead centre of the universal expansion.

After the expansion ceases and gives way to contraction, what then? Roughly speaking, it will be like running a film backwards—the film of the Big Bang and the expansion phase. The galaxies will move together and their light will appear blueshifted rather than redshifted. The 3K radio energy pervading space will 'hot up' again. Eventually the galaxies will fuse into one big agglomeration of stars. Because the distances between stars are so great, they will be unlikely to collide even in these conditions; instead they will evaporate in the intensifying heat of the collapse. Long before that stage is reached, life anywhere in the universe will be extinguished. Finally the universe will degenerate into a shrinking volume of radiation peppered with black holes which, by the simplest theory, will fuse into one conclusive point or 'singularity'.

The simplicity of the Simple Universe is almost sublime. It is wholly permeated with its own description, because its energy specifies its entire history. The same quantity may, perhaps, also specify the nature of its atoms and the forces at work within the atoms and between them, so that in a universe of different total energy matter might be subtly or radically different. Knowledge of the connection between the total energy of spacetime and the particles and forces of the universe is still too sketchy for anything definite to be said about that idea.

The fact that a single quantity, energy, describes the entire history and future of spacetime in the Simple Universe means that, in a mathematical sense, the universe is like a single dot on a piece of paper. All that happens to the universe, in the past and future and out to vast distances, is in that sense fixed. The space of the universe is certainly dot-like if you look at it from the point of view of travellers going at the speed of light. They can traverse it in zero time, because time stands still when you travel so rapidly, and the vast cosmos is completely foreshortened in any direction. As I have remarked before, it is only the existence of slow-moving matter—frozen energy—that creates measurable space and time, and gives us some room in our cosmic home to swing cats and galaxies.

Nor is it altogether impossible that the net duration of the universe is zero. The time consumed in the expansion might somehow be regarded as being recovered later, in an era of 'negative time'. Some theorists contemplate the possibility that time runs backwards during the contraction towards the Big Crunch. If so, as Michael Berry of the University of Bristol describes it: 'Light would be absorbed by stars and emitted by eyes.' More extravagantly John Taylor of King's College, London, writes of people arising from their graves, 'ungrowing and finally being unborn'.

A reversal of the familiar direction of time is, of course, hard to credit. It cannot be dismissed out of hand because theorists do not yet know enough about time—which is, since Einstein, a problem for physics, not philosophising. Roger Penrose of Oxford has offered an argument against reversible cosmic time. He says that gravity itself fixes the direction of time by its essential tendency to clump matter together.

The present direction of time is towards ever-increasing cosmic disorder. Sugar molecules scattered at random through a cup of tea represent an inherently *later* situation than the neat sugar crystals sitting in the bowl. Life and other agencies can increase order locally (in recrystallising the sugar, for example) but only at the expense of increasing disorder or 'entropy' elsewhere. In the realm of gravity, contrary to appearances, matter clumped into stars is more disorderly than matter diffused through space. A black hole (to take our customary extreme) destroys all order and all information in objects falling into it: matter becomes completely anonymous.

Penrose argues that the clumping tendency of gravity persists whether the universe is expanding or contracting, so the direction of time remains unchanged—and as we observe it. To put it another way: the real universe, as opposed to the idealised Simple Universe, evolves in detail. As a result the contraction cannot be an exact 'playback' of the expansion, and the conditions in the Big Crunch will be more complicated than they were in the Big Bang. The lack of symmetry may be sufficient to prevent time running backwards.

Almost as strange from the scientific point of view, and another sign that we have still to grasp the real character of time, is the connection of time with energy. In the Simple Universe, the rest-energy of the Sun is in effect a promissory note for six microseconds of cosmic time. This was what I had in mind in saying that time may in some deep sense *be* energy and nothing else. The link weakens, though, if the universe is actually set to last for ever—as some leading astronomers suppose.

18 : A Choice of Histories

Theorists are unsure whether the universe will collapse or not.
A Simple Universe is 'closed' and collapses.
An 'open' universe tends to expand for ever.
Space curves differently in 'open' and 'closed' universes.
An extra cosmic force could alter the prognoses.

Albert Einstein spoiled the Simple Universe that I have just described. In a paper published in 1917, 'Cosmological Considerations on the General Theory of Relativity', he introduced into the equations a 'universal constant'. It later came to be called the 'cosmological constant' and its physical meaning is that of a force, completely unknown to science, which would counteract the effect of gravity between the galaxies—a kind of cosmic anti-gravity machine. Einstein commented: 'That term is necessary only for the purpose of making possible a quasi-static distribution of matter, as required by the fact of the small velocities of the stars.'

The small velocity of *stars* is a fact; what Einstein did not know about was the high speed of the distant galaxies. By 1915 the astronomer Vesto Slipher had found that a high proportion of 'external nebulae' were moving away from us, but it was not even certain that they were indeed great galaxies like the Milky Way. Not until the late 1920s did Hubble establish that all but the nearest galaxies were moving away, and that the farther off they were, the faster they went. An astute Belgian priest was aware of what Hubble was up to and by 1927 Georges Lemaître had already used General Relativity to describe an exploding universe. It was not quite a Simple Universe in our sense, and Alexander Friedmann in Russia had largely anticipated it. But Lemaître put the Big Bang firmly on the scientific agenda.

The two decades, 1930 to 1950, were a heyday for playing God. Many theories and variants of theories were generated, which described possible universes, most of them assuming a Big Bang. But an ingenious rival to the Big Bang theory was the Steady State theory, offered by Hermann Bondi, Thomas Gold and Fred Hoyle in 1948. They accepted Hubble's finding that the universe was expanding everywhere but nevertheless contrived to keep it unchanging in a 'steady state', by invoking the continuous creation of new matter and new galaxies. That process refilled the spaces vacated by the galaxies moving apart. The theory had philosophical charms for those who would like to evade the question of where the universe came from: it was eternal and infinite in extent. It also arranged the universe to look more or less the same to anyone, anywhere in it: you could not tell where you were, or 'when' you were, by looking at the cosmic scenery.

Big Bang versus Steady State developed into a vigorous contest. It was healthy for astronomy because it stretched techniques and ideas to the limit. Among its by-products was a thorough examination of the history of the chemical elements and how they were made in stars—not in the Big Bang, as an early version of that theory supposed. In the course of the contest cosmology was transformed from a theory-spinning branch of mathematics into a fact-seeking science.

Telescopes were barely sensitive enough to reach far enough back in time to detect any change in the density of the universe. Although Big Bang seemed to be winning on points, the fight was a close-run thing until the discovery of the 3K radio energy. That was a body blow from which Steady State never recovered. The only serious non-einsteinian contender was eliminated.

By 1968, twenty years after its birth, the Steady State theory was dead. The universe is not steady: the verdict of observation is as clear-cut as that, although arriving at it was far from simple. Radio galaxies and quasars are more densely packed at distant, early phases of the universe than near at hand, and the Big Bang glows all around—in the form of the 3K microwave background.

It does not follow that an einsteinian version must therefore be right. That would be like saying, 'The cook did not commit the crime, therefore the waiter did.' It may be that no one has yet thought of a better and truer scheme for the universe. Nevertheless, in practical terms, the demise of the Steady-State theory left cosmologically-minded astronomers free to redirect their efforts to the more subtle and even trickier question of which version of Einstein's exploding universe might be correct. A new contest developed, continuing into the late 1970s with no sign of an early decision.

Few astronomers now doubt that, as an account of the story up till now, the Simple Universe described by the equations that frightened Einstein is almost indistinguishable from the universe we inhabit. Everything points to the universe beginning with a Big Bang, expanding uniformly in all directions and gradually slowing down. The prime question is whether the expansion will slow down sufficiently to bring about the Big Crunch. That has become the chief issue in cosmology, one hundred years after Einstein's birth. Less simple versions of the universe, within the scope of Einstein's theory, can start with the Big Bang and go on expanding and cooling for ever, long after the last star has burned itself out and the last black hole has exploded. In either case, one is looking many billions of years into the future.

The traditional gloss on the law of gravity is 'What goes up must come down'. Spacecraft falsified it locally; astronomers now wonder whether it remains correct cosmically. Will the galaxies now going 'up', away from one another, eventually come 'down' again, into a monstrous *mêlée*? Will the universe fry or freeze? These questions are almost, though not precisely, the same as the rather more technical issue of whether the universe is 'open' or 'closed'. Let me explain that distinction, which has to do with the warping of space in the universe at 'large, as opposed to the local effects of its individual stars.

The Simple Universe described earlier is a 'closed' universe. If you projected an extremely powerful laser beam into space in a 'closed' universe it could eventually zoom right around behind you and singe the back of your head. You would have many billions of years in which to decide to duck (longer, perhaps, than the duration of the universe). This cosmic boomerang samples the bizarre characteristics of curved space, and it makes no difference in what direction you aim your laser beam.

Coupled with this ability of light to make a complete circuit of the universe there is a trickier point to grasp, about the distribution of the galaxies. If the galaxies were simply like fragments of a bomb hurtling outwards into ordinary, pre-existing space, there would be a definite edge to the universe, marked at any instant by the fastest galaxies. The scenery around these galaxies would be quite different from that of the slower-moving galaxies. Astronomers in the trail-blazing galaxies looking forward along the line of motion would see nothing—no galaxies ahead of them. But even among the slower-moving galaxies there would be a difference: they would see fewer galaxies ahead of them than behind them.

In the 'finite but unbounded' realm of the closed universe it is not like that at all; there is no edge to the universe in that sense. All galaxies have roughly the same number of neighbours in all directions, even though there is not an infinite number of galaxies. How can that be? For a start, we have to give up the simple bomb-fragment picture of what is going on, and concentrate on the idea that space is expanding. The distant galaxies that seem to us to be hurtling away at very high speeds are not the trail-blazers: they are moving away from us rapidly because there is a great deal of space between us and them and so the expansion of space produces the high-speed motion.

Now our astronaut sets off on an imaginary journey in search of the edge of the universe. He travels a distance greater than the computed diameter of the universe and the scenery still does not change. The galaxies just go on and on. Eventually he begins to recognise some of the galaxies: he has seen them before, other side on. Like the beam of light he has come full circle, although he was trying very hard to hold a steady course. The distortion of spacetime enables the closed universe to be folded back upon itself in a way that is impossible in ordinary geometry. Distant galaxies lying in opposite directions in space are closer together than you might imagine.

Einstein's early intuition was that the universe was 'closed' in this fashion. A closed universe is said to have 'positive' curvature of space. Light bends in the same 'inward' sense as it does towards a massive body like the Sun and the mutual gravity of the galaxies (and the other contents of the universe) tends eventually to make them run together into the Big Crunch. But the possibility exists of 'negative curvature', which is more like the situation in a centrifuge, where objects press outwards from a centre. Light bends 'outwards' in such a universe and shows no inclination to return to its starting point. The galaxies tend to go on travelling apart for ever. A universe with negative curvature is said to be 'open', and it differs from a closed universe chiefly in being infinite in extent and containing an infinite number of galaxies.

There is something else to manipulate, if you are playing God, besides the curvature of space. It is Einstein's cosmological constant, equivalent to an extra universal force. Earlier I called it a 'cosmic anti-gravity machine' but it can, if you like, reinforce gravity instead of opposing it. It can operate 'outwards' or 'inwards'. These choices complicate the cosmic possibilities. Although an infinite open universe with negative

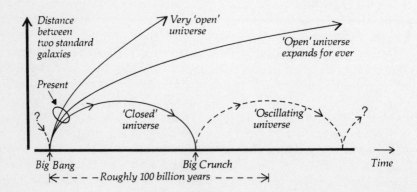

curvature of space tends to expand for ever, it can be forced to collapse if an inwards force operates with sufficient strength. Or it can shrink to a certain size and then expand again, dodging the Big Crunch. Similarly, while a finite, closed universe with positive curvature of space tends to collapse, it can be forced to expand for ever if there is an outward force of sufficient strength. An interesting case is the 'hesitating universe', where the outward force is only just sufficient to do this. The universe expands, slows down and virtually stops expanding, and then the outward force finally gets it moving again, expanding for ever.

Pure-minded relativists prefer the simple, finite, closed universe which will eventually collapse. They detest the cosmological constant and the unknown force that it represents; even though Einstein invented it he regretted doing so. Nevertheless that force may exist, after all. And, although aesthetic judgement is a powerful aid to theoretical thought, it is certainly not the final arbiter in science. Meanwhile, theorists can adjust their settings of curvature and the cosmological constant and describe all kinds of universes, with different fates, that are not inconsistent with what we know about the universe so far.

If that seems annoyingly inconclusive, then it fairly reflects the present state of the art in universe-building. The problem of deciding whether the universe will fry in the Big Crunch or freeze in the endless expansion is tricky even at the theoretical level. The task for the practical astronomer, in trying to settle the issue by observing the universe, is also formidable.

19 : Judging Fate

Astronomers are unsure whether the universe will collapse or not.
Several methods might tell whether gravity will halt the expansion.
All methods have practical or conceptual difficulties.
On balance, present evidence favours unending expansion.
A collapsing universe could give birth to a new universe.

For deciding whether the heavens will fall or not, astronomy circa 1980 is barely adequate. The most direct way to try to find out if the galaxies are going to stop in their tracks is to see at what rate they are losing speed. Somewhat less direct is the possibility of measuring the curvature of space on a large scale. Next come the attempts to estimate the density of energy, especially rest-energy or mass, in the universe: if it is high enough, gravity should be sufficient to halt the galaxies and drag them back together. Again, you can study the scattering of galaxies in space and try to infer the density. Finally, and least direct, are studies of the abundances of particular atoms in the universe, from which one can in principle deduce the energy-density of the Big Bang, and hence predict the fate of the universe. In practice even the method that is most direct in principle becomes caught up in a chain of inferences of a different kind. I shall say a little about each of the methods in turn.

Measuring the slowdown (present verdict: vacillating). The 'deceleration parameter' it is called, and the procedure is scarcely different in principle from watching a rocket fighting its way upwards in the Earth's gravity and trying to decide whether it is going to fall back to the ground or escape. The idea is simple enough. But you cannot watch any one galaxy slowing down: instead, you have to find differences between the observed speed of a distant galaxy and what you would expect if there were no slowdown. It becomes excruciatingly difficult, because the slowdown is not rapid and the distances and time-intervals involved are very great.

Measuring the speed of a distant galaxy is easy enough, using Doppler's redshift. The catch is that, at great distances, astronomers usually estimate the distance of a galaxy from the redshift, *assuming no slowdown.* If the slowdown is what you are trying to find out, you need an independent way of estimating the distances. Thus you have to resume the enterprise of Edwin Hubble, who measured the speeds of the galaxies and discovered how they increased with distance. He found himself out on an ever-longer limb of supposition about the behaviour of galaxies.

You have to assume that galaxies are much the same everywhere, so that, for example, the brightest galaxy in a distant cluster puts out just as much light, no more and no less, than the brightest galaxy in a nearby cluster. Then you can judge its distance by its apparent brightness seen from the Earth. There are at least two effects that may confound this method, by affecting the brightness of galaxies. One is that galaxies at great distances are significantly younger than nearby galaxies, because of the time taken for their light to reach us; if the galaxies were inherently brighter or dimmer at earlier stages of their evolution, the distance estimates will be unreliable. The other is a cosmic lens effect, which figures in the next method of estimating the fate of the universe and may make distant galaxies appear nearer and brighter.

Leading astronomers have done their best, but so far it is probably not good enough to measure the slowdown reliably. In 1968 the verdict seemed to be that the galaxies would halt and fall. In 1976 the consensus was the other way—that the universe was 'open' and would expand for ever—although astronomers in China were among the notable dissenters. By 1978 minds were changing again at the big American observatories, with several groups concurring that the expansion of the universe might after all be slowing down sufficiently rapidly to 'close' it and bring about the Big Crunch. A subtle use of particular emissions from quasars at great distances, proposed by Jack Baldwin of Cambridge, figured in one of the analyses. But amid such evident vacillation among the astronomers it may be prudent to treat even the latest results with caution, for the time being.

The curvature of space (no verdict yet). The Sun acts like a magnifying lens. As in the famous test of Einstein's theory of gravity, it deflects the light of stars passing near it, thereby pushing the apparent positions of the stars farther apart. (A hand lens does the same thing, pushing the ends of an insect, for instance, farther apart and making it look bigger.) In Einstein's interpretation, this lens effect is due to space being curved in the vicinity of the Sun. If the universe as a whole has positive curvature —implying a 'closed' universe and probable ultimate collapse—then empty space should act like a magnifying lens, too. The effect would be to make distant galaxies appear bigger and more luminous than they otherwise would.

Imagine a line of galaxies, all of the same size and brightness, strung out at increasing distances from the Earth. At first, the perspective is normal: the galaxies appear smaller and fainter the farther off they are.

Then, at a distance of perhaps 5000 million light-years, the magnifying effect of a closed universe (if such we inhabit) begins to operate and very distant galaxies in the line look as large and as luminous as galaxies nearer to the Earth. But the distances involved are even more formidable than for most measurements of the slowdown; and here, too, difficulties abound, associated with the evolution of galaxies. So far there has been no success with this method of judging the fate of the universe and all its contents.

The density of the universe (present verdict: no collapse). In a dense universe, with a lot of matter and energy per cubic light-year, the overall mutual gravity operating on its constituents will tend to convert an expansion into a contraction. If the universe is less dense, the speeds of the galaxies will carry them forward and, with the mutual gravity diminishing all the time, they will go on flying apart indefinitely. The decisive density is very small by earthly standards: by one estimate it needs to be roughly equivalent only to one Earth mass per thousand cubic light-years (a very large volume) to bring about the collapse of the universe. But by cosmic standards, given those yawning voids between the galaxies, that is a very high density.

The visible galaxies and their contents give a density to the universe that is only about one-thirtieth of what is needed for eventual collapse, assuming no assistance from that mysterious 'extra' force. There are invisible contributions to the density of the universe, in the form of a thin intergalactic gas, but this does not raise the density high enough to prevent eternal expansion. Other possibilities for those who would close the universe include scattering massive black holes through space and assigning mass to the swarms of particles called neutrinos which pervade the cosmos and are generally assumed to have no mass. Perhaps the most promising sign of 'missing mass' comes from the grouping of galaxies, as described now.

Clustered galaxies (present verdict: no collapse). The earlier statement that the universe seems to be remarkably simple and uniform is correct only if you are selling it in million-galaxy lots. In our vicinity it is plainly non-uniform: you and I sit on a lump of matter in a region of space dominated by the large, hot lump of matter we call the Sun. The starry environment of a galaxy such as our Milky Way is very different from the dark spaces between the galaxies. Locally, then, the universe is very lumpy indeed. In the late 1970s a group at Princeton University, led by a Canadian-born cosmologist, James Peebles, studied the scattering of galaxies in the universe. At first sight, it is completely random—as if

God has shaken his paintbrush, letting the droplets of luminous matter fall where they will. Closer examination shows a clear hierarchy of galaxies grouped into clusters, and clusters into superclusters.

A remarkably simple pattern emerged from the analysis by Peebles and his colleagues, which gave indirect evidence about the density of matter in the universe. They found that, in the real universe, clusters tend to form among galaxies lying up to about sixty million light-years apart. They mimicked the formation of clusters in a model universe, in a computer. To reproduce the observed pattern, they found that they had to assume the galaxies to be much more massive than they are generally thought to be—almost massive enough, but not quite, to halt the expansion of the universe. The problem remains of explaining why galaxies might be so much heavier than they seem to be, by conventional reckoning.

Heavy hydrogen (present verdict: no collapse). The theory of the making of the lightest elements in the Big Bang depends on well-established nuclear physics. The main products were ordinary hydrogen and helium, but there is also a significant amount of the heavy form of hydrogen (deuterium) present throughout the universe—and almost certainly produced in the Big Bang. Now, if the universe were very dense during the Big Bang there would be far less heavy hydrogen than is actually observed—it would have been 'burned' to make yet more helium.

The calculations seem quite exact and they say that the density of the universe, from the Big Bang onwards, has always been too low in relation to the speed of expansion to bring about eventual collapse. Although the reasoning may seem somewhat remote, some experts regard it as strong evidence for the fate of the universe, because it uses fairly accurate measurements of heavy-hydrogen abundances and a precise and persuasive line of argument. But the theory of the Big Bang could be defective.

The contradictory verdicts on the fate of the galaxies testify to the difficulties of the science. Those cosmologists with 'a visceral agoraphobic predilection for a closed universe' (in Martin Rees's phrase) have certainly not conceded defeat. Whether the issue will drag on for decades, or will be settled by a dramatic discovery comparable with the cosmic 3K radio energy, remains to be seen. A remarkable theoretical development is the appearance of the despised 'cosmological constant' in a completely unexpected setting—a point I shall return to later.

Although the future of the universe is still debatable, the origin of the universe and its history until now seems well described, both in evolutionary detail and in the all-embracing scheme of Einstein's equations. In half a century theorists and observers have revolutionised our perception of our place in the cosmos to a degree which matches the earlier revolution from Copernicus to Newton.

What came before the Big Bang? According to strict logic, thou shalt not ask that question. If time began with the Big Bang, the word 'before' has no meaning. But the human imagination will not be bound by logic and the question is an entirely natural one to ask. Indeed it is the point of convergence of all scientific, philosophical and religious thought. To put it bluntly: did God just say 'Let there be light!' (meaning gamma-rays) and the Big Bang ensued? Or did the energy come from 'somewhere else'? Like 'before', 'somewhere else' has no strict meaning, yet we can all sense the intention of the question.

Even the least pedantic cosmologist is, though, somewhat flummoxed by the question, because the Big Bang was such a powerful incinerator that it may have destroyed all traces of information about the 'antecedents' of all that energy. 'Wherever' it came from, the energy of the Big Bang was wholly anonymous, like the contents of a black hole. *Investigating* 'before' the Big Bang may be impossible: there can be no conventional scientific seeking of evidence, unless some new phenomenon comes to light that has survived the mangling and brought information through that dumbfounding furnace.

One solution to the problem offers itself. If ours is a Simple Universe or something like it, that will eventually recollapse into the Big Crunch, then you can quite easily imagine a new universe being born out of the ashes of the old one, in a new Big Bang . . . and so *ad infinitum*. There are technical problems about such a yo-yo universe, but it has an agreeable sort of plausibility. At the very least it gives a hint to the curious that the question of what came 'before' our universe may not be quite unanswerable. Yet the answer may be 'nothing'. As John Wheeler has remarked, it costs nothing to make a Simple Universe, in the sense that the energy put into its creation is fully recovered in its collapse. Conceivably universes are two a penny, and ours is just one among many that arise spontaneously.

To pretend that there is no religious element in this curiosity about the cosmos would be idle. I have before me the writings of a Catholic theologian who explicitly favours the interminably expanding universe, which he sees as being in keeping with 'faith in the Creator and in a creation once-and-for-all'. Hindus and Buddhists, Marxists and many agnostics would prefer the yo-yo universe: the Easterners because it accords with their idea of endless cycles, the others because they tell themselves that it removes the problem of initial creation to a comfortable distance—out of sight. These preferences are, of course, beside the point when it comes to evaluating the scientific evidence, but I mention them because they help to illuminate the passion and dedication with which scientists try to predict the long-term future of the universe. Religion as such may not come into it, but the religious urge to find meaning in life certainly does.

20 : Timeless Viewpoint

Einstein was a 'moderate' in religion and politics.
Scientifically he was dogmatic about continuity in nature.
Atomic physics deals with discontinuities in nature.
Einstein could not stomach 'uncertainty' in atomic physics.
His ideas diverged from the mainstream of discovery.

Albert Einstein himself exhibited a religious urge, without engaging in any religion. His wisecracks about God, the 'Old One', reflected a deep reverence for nature. He expressed his views succinctly in a telegram sent to a Jewish newspaper in 1929: 'I believe in Spinoza's God who reveals himself in the harmony of all that exists, but not in a God who concerns himself with the fate and action of men'.

Baruch Spinoza was the Jewish lensmaker and philosopher who reasoned that God and the material world were indistinguishable. He saw our human minds as parts of God's mind; the better you understood how the universe worked the closer you came to God. 'It is of the nature of mind,' he declared, 'to perceive things from a certain timeless point of view *(sub specie aeternitatis).'* Spinoza's writings anticipated the enterprise of modern physics and cosmology, and Einstein's regard for the seventeenth-century heretic is unsurprising. Yet even this 'soft' religiosity became, in the end, a stumbling block for Einstein the physicist.

Einstein's theories were widely reviled. A Nazi physicist described his work as 'botched-up theories consisting of some ancient knowledge and a few arbitrary additions'. The 'Jewish' theory of relativity was officially repudiated in Hitler's Germany. Hard-line religious leaders, failing to learn anything from the churches' defeats in the cases of Galileo and Darwin, also attacked Einstein. The Vatican supported an American Cardinal when he declared that relativity produced 'universal doubt about God and his creation'. And the Jewish newspaper to which Einstein sent his famous telegram about Spinoza accused Einstein of blasphemy. In the Soviet Union, Einstein's ideas were not fully acceptable in public until after the death of Stalin; for example, the *Soviet Encyclopaedia* had declared in 1925 that relativity was unacceptable from the viewpoint of dialectical materialism.

The uncomprehending antagonism evoked by Einstein's ideas is a sign that the old conflict between scientific enquiry and dogma is far from dead. Had he lived in another time or place, he might well have died at the stake, a concentration camp or a psychiatric hospital. As it was, he fled from Fascist Europe along with many of his ablest contemporaries, who were to help ensure that Hitler could not win. By harping on Einstein's Jewishness the Nazis made him a symbol of hope and pride for the persecuted Jews. Sidney Drell of Stanford has described to me the experience of young Jews in the USA and elsewhere who were inspired by Einstein's example to become scientists themselves. Nowadays a remarkable number of the world's leading scientists are Jewish by origin.

Einstein's involvement in the release of nuclear energy on Earth was painful for a former pacifist. In Berlin during the first World War, the young Einstein had courageously signed an anti-war 'Manifesto to Europeans', and then buried himself in his work. He reached the zenith of his ideas, in his theory of gravity, while the young men of Europe machine-gunned one another on every front. As he wrote to a friend: 'I quietly pursue my peaceful studies and contemplations and feel only pity and disgust'.

By 1939 Einstein was living in the USA as a refugee from the Nazis. Scientists in Berlin had just discovered the fission of uranium—quite by chance and at the worst possible moment. Nuclear physicists everywhere were quick to see both the possibility of a uranium bomb and the terrible consequences if Hitler should have it first. The problem was to persuade governments outside Germany, preoccupied with the diplomatic and military problems of the moment, to take the 'atomic bomb' seriously. Einstein was recruited as the distinguished scientist who would tell the President of the United States.

In August 1939 he signed a letter to Franklin Roosevelt, drawing the President's attention to the possibility of 'extremely powerful bombs' and warning him of evident German interest in uranium supplies. Reinforcing the urgings of others, the letter had some effect: 'This requires action,' Roosevelt said. Even so, in 1940 Einstein had to write further letters to chivvy the US government along. His own war service was as a consultant on military inventions to the US Navy. But Einstein never became a hostage to militarism.

In March 1945, he wrote to Roosevelt again, two or three weeks before the President's death, introducing to him Leo Szilard, who had drafted the 1939 letter and who wanted to urge on the President the need for international control of nuclear weapons. As the official US historians remark drily: 'This time nothing happened.' Einstein lived to see the H-bomb tests of the early 1950s letting loose doses of mc^2 far larger even than the A-bombs. So the man who had enlightened the world ended his life in the shadow of the mushroom cloud. Two days before he died, in 1955, Einstein signed a manifesto drafted by Bertrand Russell, which ended: 'We appeal, as human beings, to human beings: remember your humanity and forget the rest. If you can do so, the way lies open to a new Paradise; if you cannot, there lies before you the risk of universal death.'

Thus Einstein bobbed among the angry waves of twentieth-century life as a weatherproof prophet of reasonableness. The mutual reinforcement of science and freedom of thought has helped to shape the politics of the modern world—although that is seldom mentioned when research budgets are being negotiated. The timeless point of view is not for timeservers and it is a sovereign remedy against the virus of zealotry. Nor is it by chance, in my opinion, that one of the world's outstanding theorists of relativity, Andrei Sakharov, has become the leader of the dissidents in the Soviet Union.

If genius bears analysis at all, Einstein's seems to have been a combination of quite commonplace traits. He had a healthy scepticism about prevailing ideas, aggravated by authoritarian teaching in his schooldays. He wondered with an intense, child-like curiosity about the workings of nature and had a very good intuition concerning them. He thought visually rather than verbally. For a physicist, Einstein's mathematics was neither poor nor especially good. He enjoyed mathematical puzzles but his arithmetic was careless. He was introverted, being happier with his own thoughts than with company. And he was dogged: he tackled problems that others shrank from and, when the going was tough, he did not give up or switch to easier questions. Many people, especially scientists, have these traits in various degrees. In the young Einstein they blended into a mixture of extraordinary potency.

Wonder is the beginning of philosophy, Aristotle said. A leading historian of science, Gerald Holton of Harvard, believes that the imaginations of great scientists are formed very early in their lives. He cites as an example Einstein's own fascination with a magnet at the age of four or five, which committed him to the study of 'fields' of influence. One of the characteristics of a field of magnetism or gravity is that it seems to be continuous through space, gradually diminishing away from the source of the influence. His regard for continuity in nature served Einstein well when he was working the intellectual miracle of relativity. Unfortunately this intuitive preference conflicted with the atomic view of nature and led him to reject the implications of atomic physics—the other great emergent theme in twentieth-century natural philosophy.

He developed one appalling blind spot, incomparably more important than his self-admitted blunder of the cosmological constant, which may not after all have been a blunder. The blind spot was the quantum theory. Even though he helped to found it in its earliest years, he could not accept the quantum mechanics that evolved later and allowed 'uncertainty' to intervene in the universe. His vehement rejection of it was in part at least a consequence of his pantheistic beliefs in a perfect universe; his mild religion flawed his reason in the end. 'God does not play dice', he said, dogmatically.

'Stop telling God what to do!' Niels Bohr retorted. The Danish quantum theorist and his brilliant associates turned their backs on Einstein and showed that God does indeed play dice: he has his gaming tables in every atom and every cubic millimetre of empty space. Flowering into the theories of anti-matter, of nuclear physics, of electricity and the sub-atomic forces, quantum mechanics became a luxuriant growth, more extensive and productive than the tidier gardens of relativity. Much of our detailed understanding of the universe and life stems from it; and it is a source of new sensibility about the cosmos. The philosophically potent notion of the 'broken symmetry' asserted itself strongly in the quantum mechanics of the 1970s, notably in Steven Weinberg's description of the simple perfection that existed for a fleeting moment in the Big Bang and then 'broke', giving rise to a more complicated and more habitable material world. Einstein did not seem to realise that the perfection he craved for would be sterile.

His revulsion from the quantum theory made Einstein's own work sterile. He spent his last thirty years trying to unify gravity and electricity—but electricity could be attacked properly only in terms of the quantum theory. To be fair, the time was inopportune for what Einstein was attempting; his physicist's intuition about what was possible now failed him. Not until almost twenty years after his death was much headway made towards reconciling the theory of gravity and the quantum theory. The young Einstein might have managed something, but not Einstein past his peak. Anyway, it was not to be. Like many cocktail-party philosophers after him, the mighty Einstein fell victim to an empty pun. He confused 'uncertainty' in its sub-atomic, statistical sense with 'uncertainty' about cause and effect, which he rightly abhorred.

As individuals, scientists can be as pigheaded about their ideas as anyone else, but science itself perishes as soon as 'authority' takes charge. A continual influx of young people ready to contradict their teachers keeps it alive. The young Einstein was such a person and it would be a travesty of everything he represented to think of elevating his theories to the status of holy writ. It is true that they have serenely withstood more efforts to prove them wrong, on paper or in experiments, than any other theory in the history of science. Yet physicists now know where relativity is at risk. Even though he had plenty of confidence in his ideas, Einstein acknowledged his ultimate vulnerability. Told of a publication entitled *A Hundred Against Einstein* he remarked drily: 'One would be enough'.

21 : Einstein's Successor

The quantum theory is the key to atomic physics.
Relativity and the quantum theory disagree at short ranges.
The next big advance must reconcile them.
Present efforts include 'twistors' and 'supergravity'.
Whatever happens, Einstein's achievement is indelible.

After a famous confrontation with Niels Bohr in 1927, at a conference in Brussels, Albert Einstein retired muttering about the Old One, and Bohr emerged as the undisputed leader of theoretical physics. But the philosophical rift between Einstein and Bohr symbolised something deeper—a real gap in understanding. The theory of gravity and the quantum theory overlap without meshing. And they come into conflict at a certain point—literally a point. According to General Relativity, gravity is eventually unstoppable. Inside a black hole or a collapsing universe, the contents will tend to collapse to nothing—to a geometric point, a singularity in spacetime where a vast mass occupies zero space. Einstein himself disliked this idea, although his theory vouched for it.

Now quantum mechanics is a powerful agent for preventing collapse. It explains, for example, why ordinary atoms occupy a substantial volume of space, instead of collapsing to a dot under the influence of the electric force—which they would do by nineteenth-century theories. Sub-atomic particles occupy space by virtue of their motion, like sentries patrolling a frontier, but they are perfect sentries in the sense that they are everywhere at once. Because of a deep-rooted uncertainty about their positions the sub-atomic particles are, in effect, smeared through space. The extent of the uncertainty is not, itself, at all uncertain: it is precisely defined.

So it is with black holes and singularities. If you imagine smaller and smaller black holes you come down to a certain mass, called the Planck mass, named after the originator of the quantum theory, Max Planck. The diameter of a black hole of the Planck mass is equal to the uncertainty about its position. As a result, the black hole can no longer be pinned down in space. The Planck mass is about ten millionths of a gram. The black hole of that mass corresponds with a certain very high density of matter, such as would exist during the collapse to a singularity or at a very early stage of the Big Bang. At this distance and density physics comes to a crisis. Relativity or quantum theory (or both) must break down.

You can, if you choose, shrug off the problem as far as the singularity at the heart of a black hole is concerned: it is literally out of sight. But we can see the Big Bang—we live in its midst, in effect—and in mainstream relativity theory the Big Bang is called a 'naked singularity'. If the universe really began in a geometric point (and not everyone is convinced that it did) then physicists have no sure guide to what happened during the first moment, until the expansion reduced the density to less than that of a Planck-mass black hole. The Big Bang theory predicts that this stage is reached just 10^{-43} second (1 second divided by 1 followed by 43 zeroes) after the start of the expansion from the singularity.

An insignificant interval of ignorance? Not if you are a physicist trying to figure out the universe at its deepest levels. And even if you are sceptical about black holes and the Big Bang you cannot avoid the breakdown of physical theory—relativity in particular—at the very high densities or very short distances corresponding to the Planck mass. As a hint about what may happen at very short ranges, some theorists suspect that gravity becomes inherently far stronger—just as powerful as the electric force, in fact. That implies a story different from General Relativity.

The grand objective of physics is to understand the universe in terms as simple as possible. Looking around, the physicist sees all sorts of materials and all sorts of forces operating upon them. At first sight it is bewildering—and has been so for most of the period since the Greeks launched their formal philosophical programme of trying to make sense of the world. But by the twentieth century the progress was stunning. All ordinary materials were known to be made of atoms, and these in turn were made from a small catalogue of sub-atomic particles, mass-produced by nature. There was a very short list of cosmic forces: gravity, the electric force (including magnetism), the strong nuclear force which binds the pieces of the atomic nucleus together, and the so-called weak force which plays a key part in changing one kind of sub-atomic particle into another. Given these constituents and forces, you could explain everything.

The problem then became: 'Why?' For what reason has our universe adopted these particular catalogues of particles and forces, rather than any others? How are they related to one another and to the fundamental nature of space, time and cosmic history? The great divide between relativity and quantum theory has been the biggest obstacle to answering these questions.

A first big step towards the merging of the theory of gravity with the rest of physics came in 1974 with Stephen Hawking's theoretical discovery of exploding black holes. Very strong gravity could in effect squeeze particles of matter out of empty space, in accordance with the quantum theory. After the famous conference at the Rutherford Laboratory in England, where Hawking announced his theory, an eminent American relativist commented: 'Farewell geometry'. By that, John Wheeler meant that Einstein's notions of space were inadequate for the forthcoming reconciliation of gravity with the quantum theory, because the laws of geometry itself must fail at very short ranges.

Space becomes 'foamy' and unpredictable. According to Roger Penrose of Oxford, the breakdown of geometry begins at microscopic distances much greater than those corresponding with the Planck mass— at about the diameter of an atomic nucleus. The search for something deeper than geometry has led Penrose to embark upon his 'twistors' project. It is a theoretical programme for replacing the lines of conventional spacetime by the tracks of mathematical entities behaving somewhat like sub-atomic particles. These are the 'twistors', which possess energy and spin. With them Penrose can map out conventional spacetime, and he can also make conventional sub-atomic particles from combinations of twistors; they include the particles that carry the cosmic forces. It is a very promising project, which starts from the standpoint of a leading theorist of relativity.

Approaching the problem from the other side of the great divide are the theorists of particle physics. Their biggest single problem is a commonplace pitfall in which their calculations predict infinitely strong forces between particles. Evading that pitfall becomes a guide to how nature works. Many clever and dedicated men have been associated with this search for the rules of the cosmic game. I hope it is not invidious to mention particularly a Dutchman, Peter van Nieuwenhuizen, an Italian, Sergio Ferrara, and an American, Daniel Freedman. They produced, in nice time for the Einstein centenary, the theory of 'supergravity'. It is another impressive-looking link between Einstein's gravity and the particles of the quantum theory. The character of the theory has to do with its treatment of the particles, which brings to an end a sharp class distinction between the particles that prevailed until the late 1970s.

On the one hand there are the durable constituents of matter, principally electrons and quarks. They can change their form into other, more energetic particles of related kinds, but nature plainly keeps very careful accounts of how many of them there are altogether. If you make an electron you must make an anti-electron, so that the cosmic ledger remains tidy. Another very important characteristic of these durable particles is that they occupy space, with the result that atoms containing many electrons are physically bigger than atoms containing only a few electrons.

Apparently quite different are the force-carrying particles, like particles of light, associated with electricity, or the 'mesons' that carry the strong nuclear force. They are mixtures of matter and anti-matter—for example, a particle of light can be thought of as an electron and anti-electron working in partnership. Nature is not fussy about them: you can create particles of light any time you like just by striking a match. That makes them very suitable to serve as force-carrying particles, appearing whenever they are required. Moreover, many force-carrying particles can be crammed into the same space without limit.

In the theory of supergravity, and contrary to previous expectations, these two sorts of particles and the force-carrying particles are all variants of one another: in principle you can transform any of them into any other, by a series of steps. This might lead to universal chaos, were there not strict rules governing what particles can exist. The emergent rules turn out to be intimately related with the fact that the universe has three dimensions of space and one dimension of time.

Now you can fill the universe. Start with gravity. Assign to it one force-carrying particle, the graviton, and (say) eight novel durable particles called 'gravitinos'. Then all these particles are convertible into other particles: as a matter of fact, the theory allows for fifty-six other forms of durable particles (quarks, electrons and the like) and ninety-eight other types of force-carrying particles (light, for instance). In principle these predicted particles embrace all the known forms of matter and all of the other cosmic forces. In practice the numbers do not quite fit the known particles. Whether the theory of supergravity is imperfect, or the physicists are counting their particles in the wrong way, it is too early to say.

Suppose that supergravity is correct, what are the implications for Einstein's theory of gravity? Wonderful to relate, General Relativity is wholly incorporated within the theory. This is because of the presence of the graviton, as the particle which, in the quantum-mechanical view, is the agent of gravity. As I have mentioned earlier, one can describe gravity and the deformation of spacetime in terms of mutually-interacting gravitons, and arrive at the same answers as Einstein's. But supergravity extends and elaborates Einstein's theory by introducing the 'gravitinos'. These are durable particles of matter, which have not yet been detected. They may be very light in mass and almost impossible to detect, because they would interact with other particles extremely feebly. Alternatively, they may be so heavy as to defy creation in the present generation of particle accelerators.

Perhaps the most startling outcome of supergravity is that Einstein's cosmological constant reappears unbidden in the equations. Recall that Einstein introduced it in 1917 to prevent his relativistic universe from exploding; and then bitterly regretted doing so when Hubble showed that the universe was indeed exploding. The 'cosmological constant', corresponding with a novel force in the universe, emerges from super-gravity in a manner that depends on the strength of the electric and sub-atomic forces. It implies a deep connection between the sub-atomic laws of physics and the overall configuration of the universe.

According to the theory of supergravity the new cosmic force is operating 'inwards', creating a universe of finite size which will collapse. As we have seen, much of the astronomical evidence so far points the other way, to a non-collapsing universe, but the issue is not settled. Like the discrepancies in the number of particles, this is the kind of situation that makes science exciting.

A full century after Einstein's birth, the time seems ripe for another genius to succeed him, as he succeeded Maxwell and Newton. In the late 1970s physics and astronomy are brimming over with discoveries and ideas. We can no more guide incipient genius than direct a lightning stroke. But the point of connection between quantum theory and rela-tivity stands like a sharpened conductor high above the scientific land-scape, already glowing with St Elmo's fire and inviting a great discharge. It is entirely possible, of course, that the physics to come will be a team effort, like quantum mechanics, which had half a dozen sub-Einsteins rather than a lonely giant. Yet I doubt it: great syntheses are made in individual minds. In detail Einstein is already being surpassed; all that is lacking is the new comprehensive insight, to match his. That must come soon.

Einstein's achievement will stand. Some of his premises and preju-dices will no doubt be relegated as historical curiosities, when they are either contradicted or superseded by a greater edifice. But nothing will ever eradicate the accomplishments of 1905–17. Then, like a child play-ing with so many shining beads, he strung together matter and energy and space and time, fashioning from them a girdle for the universe. Even in our cynical century we can safely speak of the young Einstein in the words that the physicist Edmund Halley composed for his friend Isaac Newton:

'Nearer the gods no mortal may approach.'

Index

aberration, 92–94
acceleration
  equivalent to gravity, 75, 76, 78–82, 113
  in the 'twin paradox', 89, 113
accelerators, particle, 14, 15, 19, 90, 91, 98
Achilles paradox, 77, 78
Aldrin, E. ('Buzz'), 81
Alley, Carroll, 38, 81
Annalen der Physik, 11
anti-matter, 19, 20
anti-particle, 19, 20
Arecibo radio telescope, 65
Argonne National Laboratory, 70, 71
atomic clocks. *See* clocks, atomic.

Baldwin, Jack, 133
Bell Telephone Laboratories, 119
Berry, Michael, 125
Betelgeuse (star), 112, 113
Big Bang, 20, 52, 118–129, 135, 136, 143, 144
Big Crunch, 123–126, 128–131, 133
binary pulsar, 65–66, 72, 73
black hole(s)
  as a cause of curved space, 45
  collision between, 73
  detection of, 68
  doubts on existence of, 26

black hole *(cont.'d)*
  effects on light, 40–42, 55
  effects on light-bubble, 47, 48
  effects on time, 40–44, 84, 85
  'event horizon', 40
  exploding, 52, 145
  imaginary, 39, 40, 48, 63, 64, 90
  light-cone of, 114
  made from star, 24
  observation by astronomers, 4–6
  of Planck mass, 143, 144
  rotating, 22, 48, 49, 121
  small, 52, 143
  supermassive, 4–6, 24, 25, 70, 73
blueshift, 9–11
  of signals from Earth, 85, 89
Bohr, Niels, 28, 141, 143
Boksenberg, Alec, 3, 5, 6, 43
Bondi, Hermann, 127
Bradley, James, 92
Braginsky, Igor, 71, 72, 79
Brans, Carl, 80
Brans-Dicke theory of gravity, 80–82
Brecher, Kenneth, 26, 103, 104
Bristol, University of, 125

California Institute of Technology, 4, 6
centrifuge, 113
CERN, 91

149